THE BEAK OF THE FINCH

Jonathan Weiner is the author of *The Next One Hundred Years*, *Planet Earth* and *The Beak of the Finch*. He was born in New York City in 1953 and now lives with his wife and two children in Bucks County, Pennsylvania.

Jonathan Weiner

THE BEAK
OF THE FINCH

A Story of Evolution in Our Time

VINTAGE

Published by Vintage 1995

2 4 6 8 10 9 7 5 3 1

First published in Great Britain by
Jonathan Cape Ltd, 1994

Vintage
Random House, 20 Vauxhall Bridge Road, London SW1V 2SA

Random House Australia (Pty) Limited
20 Alfred Street, Milsons Point, Sydney
New South Wales 2061, Australia

Random House New Zealand Limited
18 Poland Road, Glenfield,
Auckland 10, New Zealand

Random House South Africa (Pty) Limited
PO Box 337, Bergvlei, South Africa

Random House UK Limited Reg. No. 954009

A CIP catalogue record for this book
is available from the British Library

ISBN 0 09 946871 9

Papers used by Random House UK Ltd are natural, recyclable products made from wood grown in sustainable forests. The manufacturing processes conform to the environmental regulations of the country of origin

Printed and bound in Great Britain by
Cox & Wyman, Reading, Berkshire

For Deborah

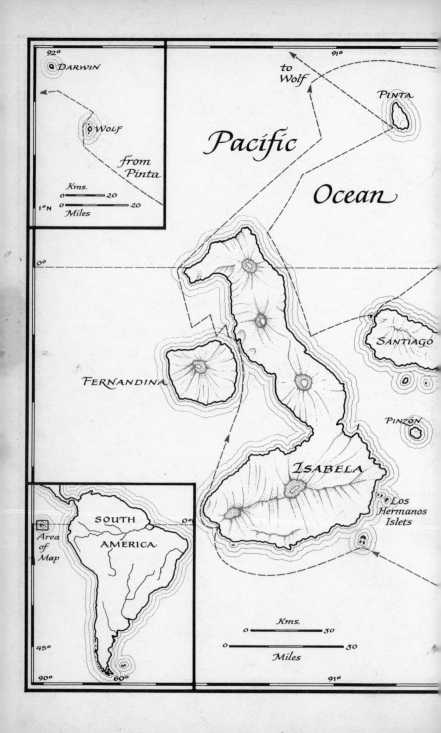

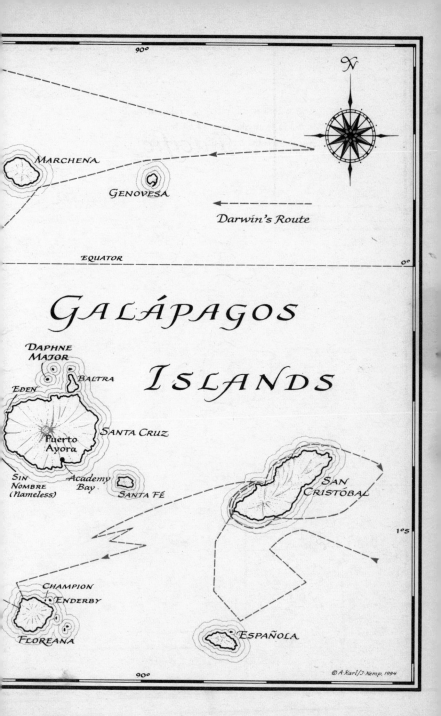

And where is the place of understanding?
It is hid from the eyes of all living;
And concealed from the birds of the air.

—*Job 28:20–21*

Contents

PART THREE: G.O.D.

PART ONE

Evolution in the Flesh

As we have heard, so have we seen. . . .

—*Psalm 48:8*

Chapter 1

Daphne Major

The Creation is never over. It had a beginning but it has no ending. Creation is always busy making new scenes, new things, and new Worlds.

— IMMANUEL KANT,
*A General Natural History
of the Heavens*

Half past seven on Daphne Major. Peter and Rosemary Grant sit themselves down on stones, a few steps from their traps. Peter opens a yellow notebook with waterproof pages. "Okay," he says. "Today is the twenty-fifth."

It is the twenty-fifth of January, 1991. There are four hundred finches on the island at this moment, and the Grants know every one of the birds on sight, the way shepherds can tell every sheep in their flocks. In other years there have been more than a thousand finches on Daphne Major, and Peter and Rosemary could still recognize each one. The flock was down to three hundred once. The number is falling toward that now. The birds have gotten less than a fifth of an inch of rain in the last forty-four months: in 1,320 days, 5 millimeters of rain.

The Grants, and the Grants' young daughters, and a long line of assistants, keep coming back to this desert island like sentries on a watch. They have been observing Daphne Major for almost two decades, or about twenty generations of finches. By now Peter and Rosemary Grant know many of the birds' family trees by heart—again like shepherds, or like Bible scholars, who know that Abraham begat Isaac, and Isaac begat Jacob; and Abraham also begat Jokshan, who begat Dedan, who begat Asshurim, Letushim, and Leummim.

In each generation there are always a few birds, just one or two in a hundred, that keep away from the Grants and refuse to be caught. This morning Rosemary, after a week of watching and plotting, has just captured two of the wariest, most difficult finches on the island. She caught them both in the space of a single minute, high on the island's north rim, next to a fallen cactus pad, in black box traps baited with green bananas. "How about that," she cried, when the traps' doors clicked shut. And when Peter strode through the cactus trees and across the lava rubble to join her, Rosemary lifted up her first prize, fluttering in a blue pouch. "I deserve a bottle of wine for this!"

Now the Grants are sitting beside the traps at the edge of a cliff, 100 meters above the Pacific Ocean. Except for the honking and whistling of two masked boobies, courting on a rock nearby, the scene is quiet. The ocean is more than pacific; it is flat as a pond. The morning's weather is what Charles Darwin described in his diary when he first saw the Galápagos archipelago, "a steady, gentle breeze of wind & gloomy sky."

From the upper rim of Daphne Major, on clearer mornings than this one, Rosemary and Peter can see the island of Santiago, where Darwin camped for nine days. They can also see the island of Isabela, where Darwin spent one day. They can make out more than a dozen other islands and black lava ruins that Darwin never had a chance to visit, including an islet known officially as Sin Nombre (that is, Nameless) and another black speck called Eden.

"If I have seen further," Isaac Newton once wrote, with celebrated modesty, "it is by standing upon the shoulders of Giants." The dark volcanoes of the Galápagos are Darwin's shoulders. These islands meant more to him than any other stop in his five-year voyage around the world. "Origin of all my views," he called them once—the origin of the *Origin of Species*. The Grants are doing what Darwin could not do, going back to the Galápagos year after year; and the Grants are seeing there what Darwin did not imagine could be seen at all.

Rosemary unlatches their tool kit, a tackle box. From it, Peter extracts a pair of jeweler's spectacles, a plastic mask with bulging lenses, which make him look like Robinson Crusoe from Mars. "Okay, Famous Bird," Peter says. "*Ow!* Famous Bird has decided to bite the hand that feeds him." He grasps the finch with one hand, and its head sticks out observantly from his fist. The bird is about the size of a sparrow, and jet-black, with a black beak and shiny dark eyes.

Rosemary hands Peter a pair of calipers. "Now, here we go," Peter says. "Wing length, 72 millimeters."

Rosemary jots the number in the yellow notebook.

"Tarsus length, 21.5." (The tarsus is the bird's leg.)

Rosemary writes it down.

"Beak length, 14.9 millimeters," Peter recites. "Beak depth, 8.8. Beak width, 8 millimeters."

"Black Five plumage." The Grants rate the birds' plumage from zero, which is brown, to five, totally black. Black Five means a mature male.

"Beak black." Normally these birds' beaks are pale, the color of horn. A black beak means the bird is ready to mate.

Peter dangles the bird in a little weighing cup. "Weight, 22.2 grams."

"This bird has lived a long time," he muses. "Thirteen years." There are only three others of its generation still alive on the island, and none older. "But I don't think there's a single one of his offspring flying around. Not *one* has made it to the breeding season." The bird has been a father many times, and never once a grandfather.

Cactus finches. From Charles Darwin, *The Zoology of the Voyage of H.M.S. Beagle.*
The Smithsonian Institution

Peter puts a gray ring and a brown ring on the bird's left ankle. He puts a light green ring over a metal one on its right ankle. Bands like these, and an ingenious color code, help the Grant team to keep track of their flocks from dawn to dusk, from the cliffs at the base of the island to this guano-painted rubble at the rim.

Peter holds the bird in his fist one more time and inspects its beak in profile. In rushing up to join Rosemary at the rim, he has forgotten his camera. Otherwise he would photograph the bird just so, from a distance of 27 centimeters. That is the Grants' standard mug shot for one of Darwin's finches.

THE ORIGIN OF SPECIES SAYS very little about the origin of species. Darwin's full title is *On the Origin of Species by Means of Natural Selection, or the Preservation of Favoured Races in the Struggle for Life.* Yet the book does not document the origin of a single species, or a single case of natural selection, or the preservation of one favored race in the struggle for life.

Darwin talks about the breeding of pigeons. He talks about Malthus, fossils, patterns in the geographic distribution of the world's flora and fauna. He marshals an enormous mass of evidence that evolution has happened. Yet Darwin never saw it happen, either in the Galápagos (where he spent only five weeks) or anywhere else.

"It may metaphorically be said," he writes in a famous passage, "that natural selection is daily and hourly scrutinising, throughout the world, the slightest variations; rejecting those that are bad, preserving and adding up all that are good; silently and insensibly working, *whenever and wherever opportunity offers.* . . . We see nothing of these slow changes in progress, until the hand of time has marked the lapse of ages, and then so imperfect is our view into long-past geological ages, that we see only that the forms of life are now different from what they formerly were."

That is Darwinism for Darwin. Life changes, down through the generations. The chief mechanism of change is the process that Darwin called natural selection. This process is at work right now around us, "*whenever and wherever opportunity offers,*" as Darwin emphasizes with his italics: not confined to a moment of creation in the dim past. It goes on this year as much as last year, now and forever, here and everywhere, like Newton's laws of motion. But the action and reaction are too slow to watch.

The invisibility of the process made its demonstration more difficult for Darwin, although the naturalist Thomas Henry Huxley, Darwin's self-appointed bulldog and griffin ("I am sharpening up my beak and claws in readiness," he wrote, as the *Origin* went on sale), met critics head on. "It has been urged, for instance, that in his chapters on the struggle for existence and on natural selection, Mr. Darwin does not so much prove that natural selection does occur, as that it must occur," Huxley wrote; "but, in fact, no other sort of demonstration is attainable. A race does not attract our attention in Nature until it has, in all probability, existed for a considerable time, and then it is too late to inquire into the conditions of its origin."

Huxley gave a public lecture with the title "The Demonstrative Evidence of Evolution." His evidence was a series of extinct ancestors of the modern horse, beginning with *Eohippus*, the "dawn horse," now called *Hyracotherium*, which lived and died about fifty million years ago. The naturalist Alfred Russel Wallace published "A Demonstration of the Origin of Species by Natural Selection," which consisted of a brief table divided in two columns. The left-hand column listed the keys to the process of natural selection. (Like Newton's laws of motion, they are so few and so simple that they can be written on the back of an envelope.) The right-hand column listed the logical consequences of these laws, ending with "changes of organic forms," or evolution. Wallace headed the items on the left "Proved Facts" and the items on the right "Necessary Consequences (afterwards taken as Proved Facts)."

Fossils argued that evolution has happened. Logic argued that natural selection can make it happen. But neither bones nor logic could demonstrate the one leading to the other, natural selection causing evolution. In 1893, in an essay entitled "The All-Sufficiency of Natural Selection," the German biologist August Weismann confessed (in italics) "that *it is really very difficult to imagine this process of natural selection in its details*; and to this day it is impossible to demonstrate it in any one point."

A few biologists did try to demonstrate it at the turn of the century. A Yankee biologist named Hermon Carey Bumpus thought he saw it at work among a flock of sparrows in Providence, Rhode Island. Other investigators reported natural selection in action among crabs in Plymouth Sound, moths in Yorkshire birch trees, mice in sandhills on an island in Dublin Bay, and chicks in a Long Island poultry yard. But most of these sitings were brief and ambiguous (Bumpus's data base was a single snowstorm). The work tended to be neglected by both sides of the debate.

Mountains of books and papers, technical and popular, were published about evolutionary theory. Much of this literature approached the level of abstraction of the medieval scholiasts' angels-on-the-head-of-a-pin. Some of the most learned interpretations of Darwinism were more or less unconstrained by reality. In crucial ways, for all the mountainous literature, the theory of evolution by natural selection was still a proof on the back of an envelope, and the origin of species remained what Darwin called, in his journal of the *Beagle* voyage, "that mystery of mysteries."

"If ever an idea cried and begged" for an experimental research program, a geneticist lamented in 1934, "surely it is this one . . . but there have been so very, very few of them." A quarter-century later, in 1960, another geneticist wrote that "the amount of observation or experiment so far carried out upon evolution in wild populations" was still "surprisingly small." He found this impoverished state of affairs disturbing because "evolution is the fundamental problem of biology while observation and experiment are the fundamental tools of science." In 1990, in a one-volume *Encyclopedia of Evolution*, a physical anthropologist wrote that the "complaint of a half-century ago holds good: The number of experimental tests of natural selection is pitiful; the few that have been conducted still do heavy duty as exemplars."

This is also the burden of the Creationists' cry, "Only a theory." According to a little paperback entitled *The Handy-Dandy Evolution Refuter*, whose cover bears the gold seal of the Chapel of the Air, in Wheaton, Illinois, "Neither evolution nor creation can be tested as a scientific theory, so believers in evolution or creation must accept either view by faith." Duane Gish, the most prominent Creationist writer today, declares in his book *Evolution? The Fossils Say No!,* "By creation we mean the bringing into being by a supernatural Creator of the basic kinds of plants and animals by the process of sudden, or fiat, creation. We do not know how the Creator created, what processes He used, *for he used processes which are not now operating anywhere in the natural universe.*" (The italics are his.)

Today more and more evolutionists are doing what Darwin thought impossible. They are studying the evolutionary process not through fossils but directly, in real time, in the wild: evolution in the flesh. "Evolution" comes from the Latin *evolutio*, an unrolling, unfolding, opening. Biologists are observing year by year and sometimes even day by day or hour by hour details of life's unrolling and opening, right now.

So many new studies are coming out that one investigator has published a technical guide for evolution watchers, a detailed and rigorous book entitled *Natural Selection in the Wild*. The centerpiece of the book is a table, "Direct Demonstrations of Natural Selection." This table begins to supply what Darwin, Huxley, Wallace, and Weismann never could. It lists more than 140 instances in which a piece of the Darwinian process has been documented. Some of these case studies, like Bumpus's sparrows, are only flashes in the storm, glimpses of the process that is at work around us, but many of the latest studies, like the Grants', are remarkably, almost panoramically complete.

Taken together, these new studies suggest that Darwin did not know the strength of his own theory. He vastly underestimated the power of natural selection. Its action is neither rare nor slow. It leads to evolution daily and hourly, all around us, and we can watch.

The Grants are leaders of this field, and they are among its ideal representatives. Year after year they go back to the most celebrated place in the study of evolution, the place that helped lead the young Darwin to his theory: the Galápagos, the Enchanted Islands. There they observe Darwin's finches, the birds that Darwin was the first naturalist to collect; the birds whose beaks inspired his first veiled hints about his revolutionary theory; the birds whose portraits in textbooks and encyclopedias have now introduced so many generations to Darwinism that they have become international symbols of the process, totems of evolution, like the overshot brows and cumulous beard of Darwin himself. Now the Grants' work on Darwin's finches is entering the textbooks too. This is one of the most intensive and valuable animal studies ever conducted in the wild; zoologists and evolutionists already regard it as a classic. It is the best and most detailed demonstration to date of the power of Darwin's process.

TO STUDY THE EVOLUTION of life through many generations you need an isolated population, one that is not going to run away, one that cannot easily mix and mate with others and, by mixing, mingle the changes induced in one place with the changes induced in others. If you detect a change in the wingspan of a bird, the teeth of a bear, the fins of a fish, or the mandibles of an ant, you want to be able to explain why the change occurred. You want to know the action to which the change is a reaction. For this you need something in nature approximating the simplicity and isolation of a laboratory.

Islands are ideal for this purpose, because it is hard for your subjects to leave them, and it is hard for outside influences to invade. Islands are like castles, communities with moats around them. Evolutionists are now watching life evolve on Gotland, in the Baltic Sea; on Mandarte, in the Georgia Strait of British Columbia; on Trinidad, in the West Indies; on the Big Island of Hawaii, in the center of the Pacific. But of all the islands in the world the nearest approach to paradise for evolutionists is still the Galápagos archipelago.

There are about a dozen major and a dozen minor islands in the Galápagos. They are the tips of volcanoes that erupted from the floor of the sea. They broke the surface of the Pacific within the last five million years or so, which makes them far younger than most of the rock that composes the continents. In fact a few of these islands are still in their birth throes, among the most fiery volcanoes on the planet. Because they are so young, the creation of new forms from old is still in the early stages in the Galápagos: life is evolving as fast and furiously as the volcanoes. And because much of this life is trapped on separate islands—the summit of each volcano is a prison for most of the creatures that live and die there—and because there was never any bridge to the mainland (South America is a thousand kilometers, or six hundred miles, to the east), the life-forms of this archipelago are following strange paths of their own.

Daphne Major, where the Grants have spent most of their time, is small and lonely even by the standards of the Galápagos. There is only one way onto the island. The Grants and their team have to go there at low tide, as early in the morning as they can, while the sea is still relatively calm, and sail around the island's base to a certain point on the south side. They can't land a boat, because Daphne Major has no shore, nothing at the waterline all the way around but cliffs two and three stories high. Most of these cliffs are steeper than walls, because the waves have cut them inward, so that the profile of the volcano at the waterline is overshot, like Darwin's brow. The Grants can't even anchor, because the waters around the island are absurdly deep, one thousand fathoms of sharks to the bottom of the ocean.

They have to leave their boat's captain describing figure eights offshore while they search along the south side of the island in a rowboat, which in the Spanish slang of the Galápagos fishermen is called a

Daphne Major, in the center of the Galápagos archipelago.
Drawing by Thalia Grant.

panga. (The origin of the word is obscure, although a wooden rowboat or dinghy laboring toward black Galápagos cliffs looks frail as a leaf of corn husk, which is also a *panga.*) They watch for a place where the cliff's rim stoops toward the water and the angle of the slope becomes slightly more inviting. Just at this spot, there is a wet black ledge near the waterline. An experienced *pangero* can find it easily. At night this ledge is often haunted by sea lions, octopuses, and night herons, but by day it is guarded only by barnacles.

The first one off the *panga* has to leap when a swell lifts the boat to the top of this ledge, which has the surface area of a large welcome mat. Often the *panga* will be flying up above the welcome mat a few meters, then dropping down below the mat a few meters, or more, depending on the mood of the ocean ("miscalled Pacific," as Darwin notes in his *Beagle* diary—for it is not always as calm as it is this morning). From the *panga* the ledge seems to shoot up as high as a ceiling and then plummet as deep as a basement.

They leap onto the welcome mat and climb the little cliff, hand over hand, on rock that is dark, wet, and many-formed, much abused by the waves, until they come to an upper ledge they call the Landing. Then they form a human chain and pass up tent canvases, bamboo poles, clothes, crates of tinned soups, all of their food for the next six months, including hefty water barrels called *chimbuzos*. They cannot land without all these provisions because there is no food or water on Daphne Major. On many days the little island feels like the solar face of Mercury. The black lava gets hot enough to fry an egg (not the proverbial egg, a real one). A jerrycan of water left out in the sun at noon can come so near a boil that it is too hot to sip. Every drop they drink they have to carry up the cliff on their backs in the *chimbuzos*, and each *chimbuzo* weighs fifty kilos, or about a hundred pounds.

Everyone in the Grants' group detests landing day. "Nobody's talking science," Rosemary says.

"Or talking," says Peter.

"It is possible to see some slightly frayed tempers," says Rosemary, mildly.

Of course, the Grants chose the island partly for its inconveniences. The whole of the archipelago was discovered by human beings rather late in the heyday of global exploration. The first historical account of them dates from the sixteenth century, when the third bishop of Panama was swept off course on a mission to Peru and almost died there. (The bishop wrote not only the first but the best one-line description of the islands: "It looked as though God had caused it to rain stones.") In the next century the place became a retreat for buccaneers. By the time of Darwin's visit there were a few settlers who led "a sort of Robinson Crusoe life" in the islands, hunting the descendants of the wild pigs and goats that the buccaneers had brought there. There was even a penal colony on the island of Floreana.

But even then, not many soldiers, sailors, jailors, pirates, or whalers would have taken the trouble to climb onto this steep little rock of an island. Those who did would have needed only an hour to walk around the island's base, and twenty minutes to walk around the rim. It is unlikely that before the arrival of the Grants and their team a single human being ever actually tried to *live* there, and even though the island is located in the very center of the archipelago it was not even included in some of the earliest maps. (It may be a nameless speck on the chart made by Ambrose Cowley, the buccaneer, in 1684, but it is not on the chart made by Alonzo de Torrés, a captain in the Royal Spanish Armada, more than a century later.) Nor did Darwin himself see Daphne Major. The *Beagle* missed it by more than a dozen kilometers. The island may have been briefly visible during the voyage of HMS *Beagle* as a homely blip on the horizon. Even today, in spite of its central location, Daphne is a rare and restricted stop for the tourist cruises that now crisscross the Galápagos. The average tourist would probably fall right off the island.

On landing day, the Grants and their assistants store some of their supplies in caves above the welcome mat. But they have to lug most of their gear almost to the rim of the volcano. This is the only spot on the island where it is flat enough to pitch a tent, aside from the crater floor, which is forbidden ground because it is the nesting place of blue-footed boobies. The trail that slants up from the Landing to the camp is not very steep, but even with the sky cloudy and a wind blowing it is still hot, muggy, and full of glare. Much of the rock, solid or loose (and almost all of it is loose and broken up to some extent), is white or near white, from many long-worn coats of guano. The whitest birds on earth, masked boobies, scream and whistle and honk from their nests along the edges of the trail, or in the middle of the trail, but do not budge. Sometimes it is hard to step around the boobies without falling off the island, for the trail is narrow, the rock is loose, and the boobies are vociferous—long darting necks, long sharp beaks, and angry honks and whistles. (The *Beagle*'s captain, Robert FitzRoy, when he landed on his first Galápagos island, called it "a shore fit for Pandemonium.")

At their campsite, the Grants lash tarpaulins to bamboo poles and prop up the poles with strings tied to piles of stones. These days they use materials that can survive the vertical sun at the equator. In earlier expeditions they used ordinary tarps. The sun and the wind beat down until, according to Trevor Price, a veteran of the finch watch, the tarp was "reduced to a symbolic flag flying from half a bamboo pole. When 'whiteys' arrived in camp," Price remembers wickedly, "they got into all sorts of contortions trying to stay in its shade and prevent pinkness."

Once they have pitched camp, however, the world of the Galápagos settles around them. They can sit on the cliff edge at sunset and watch the nearer islands turn golden. They can watch Galápagos sharks patrol the Landing, and great manta rays leaping from the water, schools of dolphins, and sometimes breaching whales. Lava lizards skitter across the rocks. Owls emerge from crooks in the rocks, and so do scorpions. Some of the finch watchers hang their boots from bamboo poles to keep scorpions from crawling into the toes.

After dark, they can sit on thrones made of relics of several shipwrecks apiece and lashed together with bits of string, and read the *Origin* by candlelight. And a single black male finch sits at the top of a cactus tree giving out long, repeated whistles, very lonely and melancholy. Before going to bed they sometimes look up and see great frigatebirds like black angels silhouetted against the moon.

The limits of the island make it almost like the frame of a work of tragic art in which someone has tried to put everything of life and death in a single place, in a single piece, in a single play. The place speaks of bare necessities, these white rocks and pale rocks and streaked lava rocks all in a pile beneath a dark gray sky and climbing out from the dark blue sea, with the long scar of the trail to the crater rim. It is an island's island, with just one half-safe place to land, one dented place to camp.

The Grants and their team live and work there like those cartoon castaways who squat on a single lump not much larger than a Galápagos tortoise, with one palm tree growing from the center. Only here, there is not even a palm tree, and the castaways are all business and vigor and eagerness, with not much time to talk.

The whole island is a diagram of limits. If a castle describes the impossibility of assault, an Alcatraz or Devil's Island the impossibility of escape, then Daphne Major suggests the near impossibility of life, and the near impossibility of its study by human beings. Yet both have triumphed. The bizarre flora and fauna hang on here drought after drought, deluge after deluge. And these biologists, all of them, team after team, year after year after year, are coming away with gold, so that the prison has become a treasure-house.

"LET'S GET ON with the measuring, darling, because this bird is a breeder too," says Peter Grant.

The beak of the bird in the second trap is as black as the first, but slightly larger. It is 15.8 millimeters long, 9.7 millimeters deep, and 9 millimeters wide. This finch is also heavier than the first, by 2.2 grams. "Probably had lots of banana," Rosemary jokes.

She and Peter band the bird's left leg with orange over black ("It's a Princeton bird, isn't it?" Rosemary says), and they band his right leg with white over metal.

The last four years on the island have been the kind that highlight Darwin's "struggle for existence." With virtually no rain, there has been virtually no breeding—so there are virtually no naive birds to catch. Despite the Grants' nets and traps, their young assistants, and their almost unlimited interest, they have never been able to catch these two birds. Rosemary succeeded this morning only after a large investment of time. She has been coming to this spot all week. On Monday she did nothing here but watch her quarry. On Tuesday she brought up two traps and baited them, but left the doors open. On Wednesday and Thursday she kept the doors wide open, changing the bananas each morning. Now it is Friday, and she has them.

Her shorts and her pink shirt are torn, and speckled by the brown sap of the *Croton* trees, which have decorated the clothes of virtually every scientist in the Galápagos since Darwin. Her hair is so light that it would be hard to say if it is blond or gray, and her cheeks, despite years beneath the equatorial sun, have the kind of rosiness that is prized on islands halfway across the planet, the British Isles, where she was born and raised.

Peter's shirt too is splotched by the sap of the *Croton* trees. He is tall, fit, wiry, with a memorable beard. In his mid-fifties, he has just started wearing glasses. He grew up on the southern edge of London, one hour's drive from Darwin's old homestead, and as the tip of his beard turns white his resemblance to Darwin is growing almost un-canny. Of course, Darwin by this age was an invalid. (His health may have been destroyed by a tropical disease, or by his own theory, which he worked on more or less in secret for twenty years after the voyage of the *Beagle*, the daily anxiety almost killing him.) Peter strides up the volcano at a pace that would be brisk on a level in England, or New England. His bare brown legs are as well toned as the legs of an athlete of twenty. He wears a set of small black binoculars around his neck with which he can identify a bird—read its I.D.—from a dozen steps away, and he whips them up often as he walks.

Rosemary swabs the finch's wing tip with an alcohol pad to clean the skin beneath its feathers. While she swabs, she chats, rather as a doctor might while preparing a patient for an injection. Just a quick prick—the bird doesn't even seem to notice. She blots the drop of blood with the same filter paper that nurses use in hospitals with new-born babies, and presses the alcohol pad to the feathers for a moment.

When the Grants leave the island, this drop of blood and the morning's numbers will travel back with them to their other lives in Princeton, New Jersey, for analysis. There too, Rosemary and Peter work side by side. They have adjoining offices at Princeton University. Rosemary is a lecturer in the Department of Ecology and Evolution; Peter is, this year, the chairman of the department.

Their tools on Daphne are low-tech; tools have to be simple to work reliably, month after month, on a desert island that Robinson Crusoe would have laughed at. But the instruments that are trained on their results in Princeton and elsewhere are among the most sophisticated in the armamentarium of science: computers, of course, to store and analyze the decades of marching numbers; and equally powerful but more exotic machines to read the coded messages that are inscribed, as if on myriad twisted and spiraling scrolls, in every drop of bird blood. Between the numbers in the notebooks, and the sentences in the blood, the Grants and others are now reading the story of life from the outside in and from the inside out. They are watching evolution in the flesh, and evolution in the blood.

"In the distant future I see open fields for far more important researches," Darwin writes in the last pages of the *Origin*. ". . . Light will be thrown on the origin of man and his history." The study of evolution in action throws light on our origin and our history, on the silent bones of Olduvai Gorge and Koobi Fora. It also casts a new light on our tumultuous present and our destiny: for the processes that are illuminated by these studies, the processes that got us here, are in turmoil. With the conditions of life on this planet changing everywhere faster and faster, the pressures of natural selection are everywhere increasing in intensity, daily and hourly, even on islands as remote as the Galápagos. Whether or not we choose to watch, evolution is shaping us all.

This is the view of life that is opening now, for those who stand on Darwin's shoulders. They can see farther than Darwin ever dreamed, and much lies in the offing, or beyond the offing.

Chapter 2

What Darwin Saw

Steeped in fable, steeped in fate . . .

—HERMAN MELVILLE,
"The Coming Storm"

There are thirteen species of finches in the Galápagos. Some of them look so much alike that during the mating season they find it hard to tell themselves apart. Yet they are also spectacularly and peculiarly diverse.

The black cocks that Rosemary trapped this morning are cactus finches. Cactus finches do more with cactus than Plains Indians did with buffalo. They nest in cactus; they sleep in cactus; they often copulate in cactus; they drink cactus nectar; they eat cactus flowers, cactus pollen, and cactus seeds. In return they pollinate the cactus, like bees.

Two other species of Darwin's finches use tools. They pick up a twig, a cactus spine, or a leafstalk, and they trim it into shape with their beaks. Then they poke it into the bark of dead branches and pry out grubs.

One finch eats green leaves, which birds are not supposed to do. Another, the vampire finch, found chiefly on the rough, remote, cliff-walled islands of Wolf and Darwin, perches on the backs of boobies, pecks at their wings and tails, draws their blood, and drinks it. Vampires also smash boobies' eggs against rocks and drink the yolk. They even drink the blood of their own dead.

There is a vegetarian species that knows how to strip the bark off twigs into long curling ribbons like Geppetto's shavings, to get at the cambium and phloem. There are also species that perch on the backs of iguanas and rid them of ticks. The iguana invites a finch to perch by assuming a posture that makes it look like a cat that wants to be petted.

Four of Darwin's finches, from his *Journal of Researches*.
   1) Large ground finch. 2) Medium ground finch. 3) Small tree finch.
4) Warbler finch.   *The Smithsonian Institution*

The whole family tree of Darwin's finches is marked by this kind
of eccentric specialization, and each species has a beak to go with it.
Robert Bowman, an evolutionist who studied the finches before the
Grants, once drew a chart comparing the birds' beaks to different
kinds of pliers. Cactus finches carry a heavy-duty lineman's pliers.
Other species carry analogues of the high-leverage diagonal pliers, the
long chain-nose pliers, the parrot-head gripping pliers, the curved
needle-nose pliers, and the straight needle-nose pliers.

   That is how they are often displayed in the textbooks, as an array
of thirteen beaks, the most famous tool kit in the natural world.
Sometimes the birds are also painted in a full-length group portrait, as
thirteen pairs—male and female, black and brown—perched in the
branches of their family tree. In either style of family portrait, they
represent the mystery of mysteries in miniature, a microcosm of the
astonishing diversity of life on earth.

   The Grants are little known outside their field, but inside it their
work on these famous birds made them celebrities years ago. "The

Grants are the gurus of Darwin's finches," says William Provine, a historian of science at Cornell University. "There's no question about that. Good grief! They are known everywhere in evolutionary biology, in every country on earth where people are working on the problems of evolution."

"The history of research in the Galápagos is in itself an interesting topic," says another historian of science, Frank J. Sulloway, who has made an extensive study of it. "You are talking about one set of organisms and also, to a large extent, one set of problems. We're not talking here about the problems of physiology or endocrinology: we're talking about the problems of evolutionary biology. And as you look at the studies that have been done from generation to generation, you can see that the research has gotten incredibly more sophisticated as time goes on. The Grants are doing things that people like Swarth or Lack would only have dreamed of doing," he says, naming authors of classic finch studies that were published in the first half of this century.

"It's like the difference between an adding machine and a personal computer," Sulloway says. "You couldn't make those calculations in the twenties and thirties. You couldn't *conceive* of those calculations in the twenties and thirties. The Grants' studies have raised this whole area to a whole new level of excellence. It's really quite extraordinary."

"Peter Grant's service to biology has been extreme," says William Hamilton, an evolutionist at Oxford University. "He has shown that the most important and pervasive theory that biology has *really does work,* and that almost all of the varied and fine details of evolution that he has found occurring are understandable by this theory, and, so far, seem to need no other. . . . I think it can be claimed that the [Grant group's] work as a whole gives the most detailed unified support to the Neo-Darwinian view of evolution that the theory has yet received."

"The problem with Daphne is, it's going to destroy bird studies forever," says David Anderson, one of the Grants' former field assistants, in the tone of a young Russian novelist lamenting *War and Peace.* "No one is ever going to live up to it. Yet everyone is going to try, because they've read the Daphne study. In a sense, it's a disaster for ornithology. The island is small enough for the Grants to know all the birds, but large enough for them to get good numbers. And they've been doing it continuously since 1973! No one's ever going to do anything like it."

Season after season, Darwin's finches keep showing the Grants more. The next view is always wider than the one behind them. They

CHARLES ISLAND

CHATHAM ISLAND

WATERING PLACE

ALBEMARLE ISLAND

Views of the Galápagos. From Robert Fitzroy's
*Narrative of the Surveying Voyages of His Majesty's Ships
Adventure and Beagle.*   *The Smithsonian Institution*

look, talk, and move like people who have discovered a fountain of youth. And if their spirits ever sag for an hour, or they begin to feel almost old for an evening, alone on their lump of rock, they are carried along by the thought of where they are.

THE PLACE WHERE DARWIN first rowed ashore is one day's short sail from Daphne Major, on the island of San Cristóbal, which the English call Chatham. The landing party included Captain FitzRoy, Midshipman Philip Gidley King, and some other *Beagle* shipmates, according to Darwin's diary entry of September 17, 1835.

Darwin had joined the voyage four years before, at the age of twenty-two, as the unpaid gentleman's companion to the captain, who was very young too and worried about the loneliness of the command, the ship's previous captain having committed suicide. By the time the *Beagle* reached the Galápagos, FitzRoy had already fulfilled his chief mission, a survey of the coast of the continent of South America. Now he and his crew had set out from the coast of Peru. They were starting across the Pacific on the long way home to England.

The men in the landing party hoped to find giant tortoises. They were looking forward to roasted tortoise meat and terrapin soup. But they did not see a single tortoise at the beach on San Cristóbal. In FitzRoy's memoir of the voyage, he describes "black, dismal-looking heaps of broken lava. . . . Innumerable crabs and hideous iguanas started in every direction. . . . This first excursion had no tendency to raise our ideas of the Galápagos Islands."

Darwin had studied the sloping black shoulders of the island from the ship, and at that distance he had thought the trees were all dead. But now as he picked his way across the beach he could see that almost every plant was "both in flower & leaf." He botanized a bit and found ten different flowers, "but such insignificant, ugly little flowers" that they would have seemed more at home in the Arctic than here in the heat of the equator. Beneath the bushes he saw small birds hopping on the lava, hunting for seeds.

"The birds are Strangers to Man & think him as innocent as their countrymen the huge Tortoises," Darwin writes in his diary. "Little birds, within 3 or four feet, quietly hopped about the Bushes & were not frightened by stones being thrown at them. Mr King killed one with his hat. . . ."

That is the way Darwin's finches enter Darwin's diary—under a hat. And for the next five weeks Darwin writes about the birds just as casually. In fact he hardly mentions them. He has too many other adventures to write about. He finds whole herds of giant tortoises, and rides on the back of one. He picks up one of the iguanas ("imps of darkness") and throws it into the water, over and over; over and over, it swims straight back to him. He yanks the tail of a land iguana that is busily digging a burrow. The lizard backs up to the surface and stares at him as if to say, "What made you pull my tail?"

As he had done all along the voyage, Darwin collected diligently in the Galápagos: "Fish in Spirits of Wine," "Reptiles in Spirits of Wine," "Insects in Spirits of Wine," and so on. He also shot a total of thirty-one finches, representing nine kinds, from three of the four islands he visited, and he stowed them all away aboard the *Beagle*. (He had learned how to stuff birds from a freed black American slave, John Edmonstone, who gave cheap taxidermy lessons at the Edinburgh Museum.)

All this matters so much to the course of human thought that the historian Frank J. Sulloway spent fourteen years figuring out what happened and what did not happen in the islands, nailing down the story finch by finch. Thanks to his detective work, the episode is now not only one of the most famous but one of the best-documented turning points in the history of science.

Contrary to legend, Sulloway has shown, Darwin did not think the finches were very important. He did not even think they were all finches. The cactus finch looked to him like some kind of blackbird; other finches looked like wrens and warblers. Darwin assumed there were plenty more just like them on some part of the coast of South America where the *Beagle* had failed to stop. In other words, the very quality that makes the finches so interesting now made them look like nothing special to Darwin. Their diversity disguised their uniqueness.

Much to his later regret, Darwin stored the finch specimens from his first two islands in the same bag, and he did not bother to label which bird came from where. Since conditions on the islands seemed more or less identical, he assumed the specimens were identical too.

He did notice that the mockingbirds he shot on his second island were slightly different from the mockingbirds on the first. For that reason he took the trouble to label these specimens, and all of the other mockingbirds he caught, by place of origin. But when the vicegovernor of the islands told Darwin that the tortoises varied from island to island as well (claiming he could tell which island a tortoise came from by its shell), Darwin more or less ignored him. "I did not for some time pay sufficient attention to this statement," he confessed later, "and I had already partially mingled together the collections from two of the islands. I never dreamed that islands, about fifty or sixty miles apart, and most of them in sight of each other, formed of precisely the same rocks, placed under a quite similar climate, rising to a nearly equal height, would have been differently tenanted. . . ."

In short, Darwin was not yet an evolutionist; he was still partly a Creationist. He was on his way home to become a country parson. That was the career for which he had trained at Christ's College, Cambridge, where he studied Scripture and collected beetles. He was more interested in beetles than Scripture, but in those days a passion for nature was considered the perfect hobby for a parson.

Darwin in the Galápagos did not have Darwin's shoulders to stand on. He had to stand on the shoulders of the giants before him. A century before, the Swedish botanist Karl von Linné had tried, as a monumental act of religious devotion, to work out the relationships of all the living forms on earth. By doing so, Linné had hoped to glimpse the plan of the Creator, the meaning of life, much as saints and scholars looked for cosmic lessons in the relationships of all the verses, chapters, and books of the Hebrew and Greek Bibles.

Linné, who wrote under the Latin name Carolus Linnaeus, divided life on earth into kingdoms, kingdoms into classes, classes into orders, orders into genera, and genera into species. It was a system so beautiful and so convenient that all Western naturalists adopted it, although as they discovered more and more species they had to add categories. (Today the major headings are kingdom, phylum, class, order, family, genus, and species.)

Linnaeus's system is often drawn as a tree of life. The trunk of the tree divides near its base to form kingdoms, and each great trunk divides again and again into ever-finer branches and twigs: into species, subspecies, races, varieties, and, at last, like leaves on the twigs, individuals. We depict the order of life, in other words, as a family tree, a genealogy, in which the branches trace back to a common trunk. Every living thing is related, whether distantly or nearly, and every animal and plant shares the same ancestors at the root.

We have grown so accustomed to this view of life (since Darwin) that a diagram of the Galápagos finches on their evolutionary tree sug-

gests to us instantly a family history, with a single ancestral finch multiplying and changing generation by generation so that there are now,
for the moment, thirteen branches.

But that is not how Linnaeus himself saw his system. To him, and
to other pious naturalists of his generation, the myriad relationships
and family resemblances that Linnaeus used to bring order to nature
did not represent anything like a genealogy of descent. Rather they
represented the plan of God, who created the species in a single week,
as described in the first pages of the Hebrew Bible: "And God created
great whales . . . and every winged fowl after his kind: and God saw
that it was good."

Darwin could read the story in his copy of *Paradise Lost*, which he
carried with him on all his inland travels. Every kind of living thing
was created in that one momentous week. It is a magnificent vision,
as if the great tree of life had sprouted in an instant, breaking the
ground and reaching with every branch to the creating sky; or as if all
finches, lions, tigers, and oak trees were born pell-mell from the cor-
nucopian womb of the earth, as Milton paints it:

> Innumerous living creatures, perfect forms
> Limb'd and full grown. . . .

And all these perfect forms, being perfect, had changed little or
not at all since Creation day.

In Linnaeus's vast botanical collections he did notice many examples of local plant varieties, variations on a theme. But in his system
these varieties were not half as significant as true species. Local varieties were merely instances in which one of the Lord's created species
had come to be adapted to its particular neighborhood. By definition,
this divergence from the original type had occurred since the moment
of Creation. Thus varieties belonged to time and to our mortal earth,
whereas species were incarnations of the holy thoughts in the mind of
God during the act of Creation.

In his later years Linnaeus wondered about this metaphysical gulf between varieties and species. A few species of plants in his collections, including certain South African geraniums, seemed to have arisen by
crossing—by hybridization. Other species seemed to have been bent and
changed by the influence of a changing environment into something
more distinctive than local varieties: they were so novel that they clearly
deserved in his scheme to be called species. Linnaeus did not take this

problem much further than hints in his diary and in the later editions of his folios. But he began to wonder if not only varieties but also these species were, as he put it, "daughters of time."

While Linnaeus himself wondered about the permanence of varieties and species, his own life's work, by virtue of its very neatness, seemed to confirm it in the minds of others. To his contemporaries, Linnaeus had brought order to the riotous diversity of the natural world. He had done for life on earth what Newton had done for the stars, planets, moons, and comets in the heavens. ("God created," they said; "Linnaeus arranged.") The living things in Linnaeus's folio volumes of natural history seemed to the naturalists who came after him to be fixtures in the universe, as the stars once seemed to astronomers: never growing or changing, never aging or dying, but shining in place as they had since the day of Creation.

Not everyone subscribed to this orthodox view of life. Darwin's own grandfather Erasmus argued the contrary view that life changes from generation to generation, and that the marvelous living intricacies and adaptations we see around us were built up bit by bit, rather than minted all at once. Another who argued for what we now call evolution was the great French naturalist Lamarck. A more obscure heretic was one of Darwin's teachers, Robert E. Grant. ("No relative that I know of," says Peter.) Grant was more or less thrown out of scientific society for his belief that living forms change down through the generations.

Arguments like these were in the air during Darwin's student days at Edinburgh and Cambridge. Nevertheless, the orthodox view was so well accepted that most naturalists in Darwin's day, including Darwin, collected type specimens essentially two by two, one male and one female. The type was supposed to be the average, the representative, the *typical*, a specimen of God's thought at the moment of Creation. Every detail of every beetle had a sacred message if we could learn to read it; even the type of the lowliest worm had begun as a thought in the mind of God. The most glorious type of all, of course, was our own, as written in the book of Genesis: "God created man in his own image, in the image of God created he him."

When Darwin collected finches, mockingbirds, and tortoises in the Galápagos, it was the type he was after: the theme, not the variations. He gathered plants and animals for the *Beagle* by the same principle that Noah collected them for the ark, two by two. Darwin in the Galápagos was still half in Milton's universe.

Besides *Paradise Lost*, Darwin brought along on the voyage the first volume of Charles Lyell's *Principles of Geology*, and although his teachers in Cambridge had warned him to take the new book with a grain of salt, Darwin had devoured it. Lyell argued that although animals and plants on this planet had indeed been created by God in an instant, and never changed since, the planet itself had been changing restlessly beneath them. Earth's crust had been rising and falling, building up and eroding everywhere since its creation. At the *Beagle*'s very first stop, at St. Iago, in the Cape Verde Islands of the Atlantic, Darwin had studied the layers of coral that are exposed in the side of the island, and seen such strong evidence of gradual, geological change that he concluded more or less on the spot that Lyell was right. Thereafter everything Darwin saw in his voyage around the coast of South America confirmed and reconfirmed the then-heretical view that the earth's surface is continually created and destroyed.

To Darwin the idea that the sculpting of the earth's surface is still going on seemed new and outrageous. He was fascinated by the thought that in this sphere, small changes can accumulate with big effect. Lyell showed him that the creation and destruction of the earth's crust is measured not in days but in ages, and that the operation continues today at the same grand, slow pace as ever.

Mountains moved, rivers moved, oceans moved, but the species of life stayed ever the same. In the second volume of the *Principles* (which Darwin received by mail at a port in South America), Lyell savages Lamarck for suggesting the contrary. "It is idle to dispute about the abstract possibility of the conversion of one species into another," Lyell writes, "when there are known causes, so much more active in their nature, which must always intervene and prevent the actual accomplishment of such conversions." Just what those barriers are, Lyell does not say, but he is convinced they exist. "There are fixed limits beyond which the descendants from common parents can never deviate from a certain type."

That is why Darwin dropped the finches from two Galápagos islands into one bag. Like Linnaeus he was well aware that different local conditions can carve a species into local varieties. He and FitzRoy had already seen evidence of that in the foxes of the Falkland Islands, and Darwin thought he saw the same thing in Galápagos rats. But Darwin did not imagine that a species would split into different varieties under the near identical conditions and skies of neighboring islands; even if they had, Darwin did not imagine that such varieties would mean anything all that important.

Nine months later, the *Beagle* was on a zigzag course across the Pacific and back to England. Darwin was working on a catalogue of his ornithological specimens, including his Galápagos finches and mockingbirds, which all rode home with him in a very cramped cabin under the forecastle. A new thought struck him, and he jotted himself a note. At that moment he was working (alas for legend) on the mockingbirds.

"I have specimens from four of the larger Islands," he wrote. The mockingbirds from San Cristóbal and Isabela looked about the same to him, but the specimens from Floreana and Santiago seemed different, and each kind was found exclusively on its own island. "When I recollect, the fact that from the form of the body, shape of scales & general size, the Spaniards can at once pronounce, from which Island any Tortoise may have been brought. When I see these Islands in sight of each other, & possessed of but a scanty stock of animals, tenanted by these birds, but slightly differing in structure & filling the same place in Nature, I must suspect they are only varieties."

Only varieties. If so, they would fit comfortably within the orthodox view of life. But what if they were something more than varieties? What if the mockingbirds had been blown to the Galápagos from the coast of South America and then diverged from their ancestors, generation by generation? What if there were no limits to their divergence? What if they had diverged first into varieties, and then gone right on diverging into species, new species, each marooned on its own island?

"—If there is the slightest foundation for these remarks," Darwin wrote, "the zoology of Archipelagoes—will be well worth examining; for such facts undermine the stability of Species." Then, in a scribble that foreshadowed two decades of agonized caution, Darwin inserted a word: "would undermine the stability of Species."

DARWIN'S COLLECTIONS WERE being talked about even before he got off the boat, because he sent home letters and crates of specimens during the voyage. The *Beagle* docked in Falmouth in October 1836, and (with a little diplomatic prodding from Darwin, who was not the only explorer bearing plants and animals from the far corners of the earth) some of the world's most learned naturalists began poring over his finds, classifying them according to the system of Linnaeus.

On January 4, Darwin donated all of his Galápagos bird skins (and other trophies) to the Zoological Society of London. Within a week, specialists at the society began talking about this new treasure trove. At the very next scheduled meeting, according to the *Proceedings of the Zoological Society*, the ornithologist John Gould announced that he was particularly excited about "a series of *Ground Finches*, so peculiar in form that he was induced to regard them as constituting an entirely new group containing 14 species, and appearing to be strictly confined to the Galápagos Islands." Gould's description of Darwin's finches made the next morning's newspapers. The London *Daily Herald* mentioned "11 species of the birds brought back by Mr. Darwin from the Gallapagos [sic] Islands, all of which were new forms, none being previously known in this country."

Soon after that, Darwin took an apartment in London: he wanted to be near Gould and the other specialists who were working through his collections. In mid-March he visited Gould at the Zoological Society and asked about his Galápagos specimens. The sheet of paper on which Darwin jotted his notes at this meeting is preserved at the Cambridge University Library. Both sides of the paper are full of racing scribbles, which are themselves covered over on each side with one large scrawled word, *Galápagos*.

Gould summarized what he had learned so far about the Galápagos specimens. Almost all of the land birds were new, Gould said; they had never been described before, and apparently they lived only in the Galápagos. Three of the mockingbirds were not just local varieties, in Gould's opinion. No, as Gould had already informed the members of the Zoological Society, they were separate species. This was the verdict that Darwin had conjectured "would undermine the stability of Species."

What is more, the tame little birds that Darwin had found hopping around beneath the bushes were unique too. They were not relatives of blackbirds, warblers, wrens, and finches, as Darwin had thought when he bagged them. They were all finches, a strangely diverse group of finches, and they were all unique to the islands. Darwin squeezed in Gould's names for them on the back of his notepaper, down at the very bottom.

That was the fabulous moment—not out on the islands but indoors in a cluttered office in London. Or rather that was one in a swift series of moments, of intellectual shocks, that set Darwin reeling as expert naturalists told him more and more about his finds. The giant tortoises are unique to the Galápagos as well. So are the marine iguanas, Darwin's "imps of darkness." So are the very bushes and the cactus trees. Species after species in the Galápagos bears a family resemblance to relatives on the mainland of South America but are clearly distinct from anything ever found there. Year after year these revelations fanned the ember of Darwin's secret thought. To Darwin all these species, marooned in their lonely archipelago, had diverged from their ancestral stocks and then gone right on diverging. They had broken the species barrier.

Darwin's fossils from South America turned out to be exciting too, although while he was digging them up and crating them he had wondered if old bones were worth the work. Many of them proved to be extinct relatives of living forms. The continent of South America is the home of the armadillo, the llama, and the capybara, which is a rodent the size of a hog. Among the fossils that Darwin had found were a giant armadillo, a giant llama, and a rodent the size of a rhinoceros. These fossils helped to confirm what Lyell and other geologists had already guessed from finds in Australia. There is a "law of succession" that links the living to the dead, the same law that links the fossils of one stratum of rock to the fossils in the strata below.

If the giants he had found in the earth were ancestors of the animals he had watched on top of it, then Darwin could read in them the same thrilling story he read in the Galápagos. Whether he followed his finds horizontally, tracing the spread of animals and plants across the surface of the earth, or vertically, tracing them down into the abyss of time, the same secret stared back at him.

"It was evident that such facts as these, as well as many others, could be explained on the supposition that species gradually become modified," Darwin wrote long afterward; "and the subject haunted me." That spring he made his first jottings about evolution in a red notebook he had started on the *Beagle*. And that summer he opened his first notebook on "Transmutation of Species."

In the memoir that he worked up from his diary, the *Journal of Researches* (better known as *The Voyage of the Beagle*), he writes at some length about the birds of the Galápagos, especially the finches, "the most singular of any in the archipelago." He notes in a famous passage that "in the thirteen species of ground-finches, a nearly perfect gradation may be traced, from a beak extraordinarily thick, to one so fine, that it may be compared to that of a warbler. I very much suspect, that certain members of the series are confined to different islands. . . ."

What follows next is the first published hint of his secret theory: "Seeing this gradation and diversity of structure in one small, intimately related group of birds, one might really fancy that, from an original paucity of birds in this archipelago, one species had been taken and modified for different ends."

Then he breaks off: "But there is not space in this work, to enter on this curious subject."

Of the Galápagos as a whole, he concludes his memoir with this tantalizing and magnificent line: "Hence, both in space and time, we seem to be brought somewhat near to that great fact—that mystery of mysteries—the first appearance of new beings on this earth."

Two years after the voyage, Darwin married, and he and his wife, Emma, took a house in London. He fell ill. His stomach was seldom "right" for a whole twenty-four hours. He suffered from anxiety, boils, dizziness, eczema, flatulence, gout, headaches, insomnia, and nausea. He spent the rest of his life pursuing his researches and writings in the bosom of his growing family, first in London, then in the rural privacy of Downe, Kent.

Only after his electrifying meeting with Gould had he realized how interesting the Galápagos finches might be to him, since they are by far the most numerous and the most diverse land birds in their remarkable archipelago. He had asked permission to look over the finch skins in the collections of Captain FitzRoy and other shipmates, including his own servant on the *Beagle*, Syms Covington. The captain and the servant had labeled their finches island by island, for the ironic reason that they were not collecting the birds according to a scientific theory—they were just collecting. From their neatly labeled specimens, Darwin tried to figure out where he had caught each of his own finches—and he failed.

STILL, DARWIN DID NOT WASTE his time moaning, like Lord Jim, "What a chance missed!" Even at Down House in Kent he was able to gather enough evidence so that he could say, in the *Origin*, "The great power of this principle of selection is not hypothetical."

What Darwin developed, as a living demonstration of his theory, was an analogy. He studied the power of breeders. People had been shaping and molding animals and plants since before the time of the shepherds in the land of Israel. There are allusions to the practice in Genesis. There are treatises on animal husbandry in ancient Chinese encyclopedias. Breeders in England in Darwin's own time were spectacularly active, making new breeds of sheep and cattle, new varieties of strawberries and roses. The best of the new breeds were exported all over the world. British racehorses and bulldogs with fine pedigrees fetched high prices.

Darwin knew that breeders could shape not only animals' bodies but their very instincts. In the *Origin* he notes that "a cross with a bulldog has affected for many generations the courage and obstinacy of greyhounds; and a cross with a greyhound has given to a whole family of shepherd-dogs a tendency to hunt hares." Darwin tells of one dog "whose great-grandfather was a wolf, and this dog showed a trace of his wild parentage only in one way, by not coming in a straight line to his master, when called."

There were some who thought that God created each kind of domestic animal and plant separately; they argued "for every variety being an aboriginal creation." But Darwin knew from reading the breeders' treatises that the power lay with the breeders themselves. They called their secret the power of "picking," or "selection." Picking the most prolific hen in the chicken coop, the fleetest horse in the field, or the best rose in the garden, over and over again, according to one treatise that Darwin read, "enables the agriculturalist, not only to modify the character of his flock, but to change it altogether. It is the magician's wand, by means of which he may summon into life whatever form and mould he pleases." The results were like Creation itself. One British lord, praising the work that breeders had done with sheep, wrote, "It would seem as if they had chalked out upon a wall a form perfect in itself, and then had given it existence."

Since breeders called the art of choosing "selection," they called any changes in a breed that did not take place because of their conscious efforts—all of the casual, frustrating, and inexplicable changes in their flocks and herds behind their backs—"natural selection."

To see the selection process firsthand, Darwin took up the breeding of pigeons. In 1855, twenty years after his visit to the Galápagos Islands, Darwin began collecting pigeons in a coop in back of Down House. Though he now loathed travel, Darwin went into London to the Freemason's Tavern to meet with gentleman pigeon fanciers of the

English pouter and English
fantail. From Charles Darwin,
*The Variation of Animals and
Plants under Domestication.*
The Smithsonian Institution

Philoperisteron Society. He went to pigeon shows and poultry shows. He asked the secretary of the Philoperisteron Society, William Tegetmeier, to buy pigeons and their skeletons for him at Covent Garden.

"Very many & sincere thanks for keeping me in mind about Pigeons," he wrote to Tegetmeier on New Year's Day, 1856, and two weeks later he wrote apologetically to a neighbor, "I had meant to have sent you a line on Sunday, but quite forgot it myself. —Indeed we are all sick & miserable, & I hardly care even for Pigeons, so you may guess what a condition I am in!"

In April 1856 Darwin stood in front of his pigeon coop with the geologist Charles Lyell. By now Darwin had fifteen breeds in his coop, including tumblers, trumpeters, laughers, fantails, pouters, polands, runts, dragons, and scandaroons. These birds were so very different from one to the next, as Darwin explained to Lyell, that if they had been found in the wild they would have been classified by biologists as belonging to separate species, or even as separate genera—distinct *groups* of species. Yet all these breeds had been created by nothing more mysterious than selection. If selection could do this much with

An English carrier. From Charles Darwin, *The Variation of Animals and Plants under Domestication.*   The Smithsonian Institution

pigeons in the short space of human history, how much more could the same power do in nature in the course of millions and millions of years, in the spans of time that move mountains?

Lyell was not converted by Darwin's pigeons, but he was impressed. He urged Darwin to publish something fast, to establish priority for his ideas about evolution and natural selection. On Lyell's urging, then, Darwin began the great writing project that resulted at last in his *Origin of Species*. He called it his "Big Book." His working title was *Natural Selection*.

Darwin knew better than to hang this book on the finches he had

The rock
pigeon,
parent-form
of all Darwin's
fancy pouters,
laughers, fantails,
and scandaroons.
From Charles Darwin,
*The Variation of Animals
and Plants under Domestication.*
The Smithsonian Institution

collected so haphazardly as a young man in faraway islands. He describes the finches in *Natural Selection*, the long first draft of his manuscript, but he does not even mention them in the final draft. Instead Darwin begins the *Origin* with his pigeons, including the English carrier ("greatly elongated eyelids . . . a wide gape of mouth"), the short-faced tumbler ("with a beak in outline almost like that of a finch"), and the common tumbler ("has the singular and strictly inherited habit of flying at a great height in a compact flock, and tumbling in the air head over heels").

Generations of readers have wondered (behind drooping lids) why Darwin goes on so long about pigeons when his true theme is so much more exciting. Why talk about change in pigeon coops and hothouses when Darwin's subject is turbulence in the natural world? But farms and nurseries were the only place where Darwin had seen it happen, and the only place he thought human beings could see it.

"In Darwin's treatment of the subject, no proof is adduced that a selective process has ever been detected in nature," wrote the British evolutionists Guy C. Robson and Owain W. Richards in their influential book *The Variation of Animals in Nature*, published in 1936. "Throughout the work such a process is suggested and assumed: its actual occurrence is nowhere demonstrated. Stated briefly, the argument is as follows: selection has plainly 'worked' in domesticated races, analogous results and appropriate processes and conditions are found in nature, therefore we may assume that selection works in nature. In short, the proof is based on circumstantial rather than direct evidence, and the mainstay of the case is the analogy between Artificial and Natural Selection."

Robson and Richards add, "It is a very unsatisfactory state of affairs for biological science that a first-class theory should still dominate the field of inquiry though largely held on faith or rejected on account of prejudice."

THE FABLE THAT HAS GROWN up around Darwin and his finches is a Hollywood version of the romance of the mind. It simplifies the story by putting all those pigeons and mockingbirds in the background (Get those birds off the set!), and it speeds up the action by making it love at first sight, an Archimedean *eureka*. Darwin is so impressed by the finches and their beaks that his theory of evolution leaps straight into his head. He sails away dizzy with impious visions, as if he has just tasted an apple from the Tree of Knowledge.

In one popular account the young man sits down by one of the giant tortoises' drinking holes and stares at the gray domes of the tortoises, rather the way the young Hamlet sits in a Danish graveyard and ponders a skull. "If there are differences in the beaks of the finches and the shells of the tortoises, I must exercise extreme care to label each island's collection quite scrupulously," Darwin muses. ". . . That could be the most important discovery of my journey. What causes these differences? 'Aye, there's the rub.' "

Another popular account asks us to imagine Darwin arguing his finch theory with Captain FitzRoy in their narrow cabin, "or, if you like, out on the poop deck on a calm night as they sailed away from the Galápagos, putting forth their ideas with all the force of young men who passionately want to persuade one another and to get to the absolute truth." FitzRoy dismisses Darwin's ideas as "blasphemous rubbish," but Darwin hurls his notions back "against the blank wall of FitzRoy's uncompromising faith," as if "battering down the Church itself."

Millions of students have been taught this fable, and thousands are still taught it every year. "It has become, in fact, one of the most widely circulated legends in the history of the life sciences," writes the historian Frank J. Sulloway, "ranking with the famous stories of Newton and the apple and of Galileo's experiments at the Leaning Tower of Pisa, as a classic textbook account of the origins of modern science." Sulloway tried to break the spell on the hundredth anniversary of Darwin's death. He published not just one but a series of brilliant papers on Darwin's conversion and on the evolution of the legend. Yet the legend is still in wide circulation.

ALTHOUGH DARWIN HIMSELF never went back to the Galápagos, many of the naturalists who sailed there in his wake collected finches. In 1868, 460 specimens. In 1891, about 1,100 specimens. In 1897, 3,075 specimens. In the ambitious expedition of the California Academy of Sciences in 1905–6, 8,691 specimens. By then Darwin's finches had become one of the best-known tribes of birds on the planet.

Just by looking at these birds, three generations of biologists felt as if they could almost see evolution in action—once Darwin had opened their eyes to the process. But the Grants are the first scientists equipped with enough patience, stubbornness, ground support and sea support, enough computer power, airplane power, and staying power, to watch the process actually happen.

Chapter 3

Infinite Variety

> ... where, if we may use the expression, the manufactory of
> species has been active, we ought generally to find the
> manufactory still in action. . . .
>
> — CHARLES DARWIN,
> *On the Origin of Species*

Peter Grant began to wonder about the variation of animals and
plants during his undergraduate days at Cambridge University, the
university where Charles Darwin was a fair-to-middling divinity student. After graduating, Grant went on brooding about variation
while studying goldfinches and cardinals on the Tres Marías Islands
of Mexico, nuthatches in Turkey and Iran, chaffinches in the Canar-
ies and the Azores, mice and voles around McGill University in
Canada.

Rosemary came to the subject even younger. She grew up in a
village in the Lake District, where she lived very much outside, and
she remembers trailing after the old family gardener when she was
four years old, asking him why individual plants, birds, and people are
different one from the next. A row of vegetables, and no two were exactly alike. Birds: you could tell them apart. It was true of all the trees
roundabout, beeches, birches, oaks, and ash; and all the common
birds, tits, robins, blackbirds, finches.

"What applies to one animal will apply throughout time to all
animals—that is, if they vary—for otherwise natural selection can do
nothing," Darwin says in the *Origin*. Slight variations, in Darwin's
view, are what the process of natural selection is daily and hourly scrutinizing. Variations are the cornerstones of natural selection, the beginning of the beginning of evolution. And as Darwin shows in the first
two chapters of the *Origin* with wild-duck bones, cow and goat ud-

ders, cats with blue eyes, hairless dogs, pigeons with short beaks, and brachiopod shells, variations are everywhere.

Darwin studied the variation problem most deeply not in birds but in barnacles. In October 1846 he began trying to classify a single curious barnacle specimen that he had found on the southern coast of Chile. It was the very last of his *Beagle* specimens, an "illformed little monster," the smallest barnacle in the world. To classify that barnacle he had to compare it with others. Soon the working surfaces of his study were littered with barnacles from all the shores of the planet.

The classic barnacle is an animal with the body plan of a volcano: a cone with a crater at the top. It colonizes rocks, docks, and ships' hulls. Every day when the tide rolls in, each barnacle pokes out of its crater a long foot like a feather duster and gathers food. When the tide goes out, each barnacle pulls in the feather duster and clamps its crater closed with an operculum—a shelly lid. To mate, a barnacle sticks a long penis out of its crater and thrusts it down the crater of a neighbor. Since every barnacle in the colony is both male and female, this is not as chancy as it sounds.

What could be more of a sameness than a colony of barnacles? But

A few of Darwin's barnacles. From Charles Darwin, *A Monograph on the Sub-class Cirripedia*, volume 2.   *The Smithsonian Institution*

Darwin, staring through a simple microscope, found himself descending into a world of finely turned and infinitely variable details. He wrote to Captain FitzRoy: "for the last half-month daily hard at work in dissecting a little animal about the size of a pin's head . . . and I could spend another month, and daily see more beautiful structure."

In every barnacle genus he found astonishing variations. In one genus, "the opercular valves (usually very constant) differ wonderfully in the different species." Elsewhere he found variations in the form of "curious ear-like appendages," "horn-like projections," and, in one strange species, "the most beautiful, curved, prehensile teeth."

Everywhere he looked, individual differences graded into subspecies, subspecies shaded into varieties, varieties slid into species. Which specimens were the true species? Where should he draw the line? "After describing a set of forms as distinct species, tearing up my MS., and making them one species, tearing that up and making them separate, and then making them one again (which has happened to me), I have gnashed my teeth, cursed species, and asked what sin I had committed to be so punished."

(Darwin's friends knew just how he felt. The botanist Joseph Hooker wrote Darwin across the top of one letter, "I quite understand and sympathize with your Barnacles, they must be just like Ferns!")

This profusion and confusion of barnacles helped confirm for Darwin that one species can shade one into another: that there is no species barrier. In many cases Darwin discovered one barnacle subspecies, variety, or race (he did not know which to call them) on rocks at the southern edge of a species' range, and another subspecies, variety, or race at the northern edge of its range. In *Natural Selection*, his sprawling first draft of the *Origin of Species*, he notes that in many of these cases, "natural selection probably has come into play & according to my views is in the act of making two species."

Of course Darwin assumed that the split, the act of creation, would be much too slow to observe any motion in his lifetime, because evolution proceeds at a barnacle's pace. Darwin proceeded at a like pace through his barnacles. "I am at work at the second volume of the Cirripedia, of which creatures I am wonderfully tired," he wrote in 1852, when he had been plowing through barnacles for six years, and still had one more year to go. "I hate a Barnacle as no man ever did before, not even a sailor in a slow-sailing ship."

Variation is both universal and mysterious, one of the deepest problems in nature, and for Darwin it was for a long time completely bewildering. He wondered why, if his thinking was right, we see any species at all. Why not a continuous spectrum from tiny individual variations right on up the scale to kingdoms? Why for instance do we find a vampire finch and a vegetarian finch? (An example that Darwin might have liked, if he had known about the vampires.) Why not a whole smooth series of omnivores between the two, with a perfect series of intermediate beaks? Why not a blur, a chaos, an infinite web or Japanese fan of continuous variations?

In the sixth and final edition of the *Origin of Species*, in his chapter "Difficulties of the Theory," Darwin puts this objection at the very top of the list: "First, why, if species have descended from other species by fine gradations, do we not everywhere see innumerable transitional forms? Why is not all nature in confusion, instead of the species being, as we see them, well defined?"

Very briefly, Darwin's explanation is that the same process that makes varieties also destroys them. In the struggle for existence, some variants do better than others. When we look around us in the Galápagos or in Jersey or in New Jersey, the species of animals and plants we see are survivors. Varieties in between them have died off and disappeared, so that, after the long lapse of ages, we see only the victors and not the intermediate forms; we see the spines but not the webbing of the Japanese fan. "Thus," Darwin says in the *Origin*, "extinction and natural selection go hand in hand."

By this reasoning, if we were present at the creation of new species, if we could locate a point of origin, a place where the tree of life is growing new branches right now, we would see something less distinct and more chaotic. We would see a blur of variations shading from the individual up to the level of the species, or even up to the level of the genus. Wherever naturalists find such a blur they should suspect that here is a place where evolution is in fast action, where species are in the act of being born. Of course, Darwin thought even this fast action would be too slow to watch. We would know that we stood at the point of origin of new forms not by seeing turbulent motion, but only by seeing a sort of frozen foam from which we could infer the waterfall, "an inextricable chaos of varying and intermediate links."

The thirteen Galápagos finches are just such an inextricable chaos of varying links. Darwin never knew how many links there really are, the chaotic and almost continuous variation in his Galápagos finches, because he brought back only thirty-one specimens, but he got an inkling of the problem when he watched the experts struggle with them. At first the ornithologist John Gould named one finch *Geospiza incerta*, meaning "ground finch, I guess." Later Gould changed his mind and lumped that specimen with another. The fact that Gould ended up with the same total number of species of Galápagos finches as taxonomists currently do is a coincidence, because Gould's thirteen were not the modern thirteen.

Taxonomists can be classified into splitters and lumpers. Faced with the diversity of Darwin's finches, some splitters recognized dozens and dozens of species and subspecies. Some lumpers went so far as to call them all a single species. Generation after generation of naturalists made brief pilgrimages to the Galápagos, or puzzled over the

Darwin's ground finches.

1) Medium ground finch, *Geospiza fortis*. 2) Large ground finch, *Geospiza magnirostris*. 3) Sharp-beaked ground finch, *Geospiza difficilis*. 4) Small ground finch, *Geospiza fuliginosa*. 5) Large cactus finch, *Geospiza conirostris*. 6) Cactus finch, *Geospiza scandens*.   *Drawings by Thalia Grant*

specimens at the British Museum and the California Academy of Sciences. There were so many freaks, so many misfits that broke the serried ranks in the museum drawers. "The extraordinary variants," one ornithologist declared in 1934, "give an impression of change and experiment going on." Naturalists read and reread the reports of those who had seen Darwin's finches alive, they sorted and resorted the stiff little rows of specimens in the museums, and they wondered what on earth was happening on Darwin's islands.

Today most taxonomists consider the thirteen species a single family (some say subfamily) of birds. Within this family or subfamily, taxonomists think four groups of species are particularly closely related, and so, for the moment at least, most taxonomists divide the family of Galápagos finches into four genera. In one genus, the birds all live in trees and eat fruits and bugs. In the second genus, the birds also live in trees, but they are strict vegetarians. In the third genus, the birds live in trees, but they look and act like warblers. In the fourth genus, the birds spend most of their time hopping on the ground.

This last group is the largest, with six species. It is also, for obvious reasons, the easiest to watch, and from the beginning the Grants and their team have focused on it. The Latin name of this genus is *Geospiza*: ground finches. The ground finches are a strange little club in themselves, a microcosm within a microcosm. The membership list includes the sharp-beaked ground finch, *G. difficilis*; the cactus finch, *G. scandens*; and also a large cactus finch, *G. conirostris*. Then comes a trio that has become as familiar to the Grants as Goldilocks' Three Bears. There is a large ground finch, *G. magnirostris*; a medium ground finch, *G. fortis*; and a small ground finch, *G. fuliginosa*. The large ground finch has a large beak, the medium ground finch has a medium-sized beak, and the small ground finch has a small beak.

Within each of these three species, the beaks of individual birds are variable. That is, they blur together, just as we would expect if this is a place where the river is racing and the cosmic mills are turning fast. For instance, the species in the middle of the trio, the medium ground finch, *fortis*, sometimes shades into the species above it, *magnirostris*, or the species below it, *fuliginosa*. The very biggest specimens of *fortis* are just as big as the very smallest specimens of *magnirostris*, and so are their beaks. At the same time the very smallest specimens of *fortis* are just as small as the biggest *fuliginosa*, and so are their beaks.

Some of the world's biggest *fortis* live on the island of Isabela; some of the world's smallest *magnirostris* live on the neighboring island of

Rábida. The largest of the *fortis* on Isabela are, even to Peter and Rosemary Grant, "almost indistinguishable" from the smallest of the *magnirostris* on Rábida.

You can't distinguish these three species by their plumage, and usually not by their build or body size either. You have to tell them apart by their beaks. In the jargon of taxonomy, the sullen art of classification, the beak of the ground finch is diagnostic: it is the birds' chief taxonomic character. But because the finches and their beaks are so variable, many of them "are so intermediate in appearance that they cannot safely be identified—a truly remarkable state of affairs," as the ornithologist David Lack sums up in his famous monograph, *Darwin's Finches*. "In no other birds are the differences between species so ill-defined."

"CAUTION," says a modern field guide to the birds of the Galápagos: "It is only a very wise man or a fool who thinks that he is able to identify all the finches which he sees." At the Charles Darwin Research Station, on the island of Santa Cruz, the staff has a saying: "Only God and Peter Grant can recognize Darwin's finches."

PETER GRANT, having studied his chaffinches, nuthatches, mice, and voles, wondered what makes some species of animals and plants hypervariable and others not. He wondered why some of the most variable species will vary even in their variability, with one flock full of eccentrics and another flock full of conformists.

These were outstanding biological questions in the early 1970s, when Grant began looking for his next research project. (At the time, Peter was in charge of the research; Rosemary was in charge of logistics.) Theoretical and mathematical biologists were advancing paper dragons at one another in the pages of learned journals. Grant wanted to watch what actually goes on in nature. What he needed was a group of hypervariable species, well studied, variably variable, scattered across a set of remote and undisturbed locations. "The Galápagos were ideal," he says. "Darwin's finches were *ideal*."

The Grants made their first trip to the Galápagos in 1973. That first year they worked with one of Peter's postdoctoral students, Ian Abbott, and his wife, Lynette, among others. To the bemusement of Peter's and Rosemary's families in England, the Grants also brought their two daughters, Nicola and Thalia, then aged eight and six. The girls had already camped with them in Greece, Turkey, and Yugoslavia to watch nuthatches. (In those young field days, Rosemary's biggest job was catching Nicola and Thalia.)

They found the birds as tame as in the days of Darwin, or of the shipwrecked bishop Berlanga, in 1535, who marveled at the birds "which did not fly from us but allowed themselves to be taken." This is one of the strangest things about the Galápagos, after the strangeness of the animals themselves, and almost everyone who has ever written about the islands exclaims at it, including Cowley the buccaneer and Lord Byron (successor to the poet), who stopped there while returning a dead Polynesian king and queen to the Sandwich Islands.

Cowley wrote, in 1699, "Here are also abundance of Fowls, viz., Flemingoes and Turtle Doves; the latter whereof were so tame, that they would often alight upon our Hats and Arms, so as that we could take them alive, they not fearing Man, until such time as some of our Company did fire at them, whereby they were rendered more shy."

Byron wrote, in 1826, "The place is like a new creation; the birds and beasts do not get out of our way; the pelicans and sea-lions look in our faces as if we had no right to intrude on their solitude; the small birds are so tame that they hop upon our feet; and all this amidst volcanoes which are burning round us on either hand."

"The finches are much more afraid of hawks and owls than they are of us," Peter Grant tells friends in Princeton. "When we walk up to them, the birds keep doing what they are doing; but when an owl comes near, they head for a cactus tree. A little while ago Rosemary was crossing a treeless spot. An owl flew over, and finches flew up from all around and landed on Rosemary!

"They are always perching on our shoulders, our arms, our heads. Sometimes when I am measuring one, a few others land up and down my wrists and arms to watch. I was looking out to sea with binoculars once, and a hawk landed on my hat. We have a picture of it."

"Or you pick up a bamboo pole, set it over your shoulder, and begin to carry it," says Rosemary. "Suddenly the pole is very hard to hold up. You are walking along wondering why it is so heavy. Then you turn around and see that you have a hawk hitching a ride on the back of it."

"I used to have a wart on my back, although it is gone now," says Peter. "A small black wart, up near the right shoulder. I went around in just shorts in those days, and on Genovesa, finches would peck at the wart."

"What a difference!" says one veteran of the finch watch, Dolph

Galápagos hawks. From
Charles Darwin, *The
Zoology of the Voyage of
H.M.S. Beagle.*
The Smithsonian Institution

Schluter. (Schluter coined some of the team's favorite names for itself, including *El Grupo Grant,* the Finch Unit, and, for maximum grandeur, the International Finch Investigation Unit.) "In Kenya, finches flush as much as 30 meters away. In the Galápagos, the birds land on the rim of your coffee cup. If there's only a little coffee left in there, they will land right inside it and take a sip. You can put your hand over it, and measure the bird. Mockingbirds on Genovesa would pick at our shoelaces. On the really isolated islands like Wolf, you could catch the birds by hand. Just reach out your hand and grab them."

Peter, camping in Shiraz, had once spotted a pair of nuthatches feeding near a rock. He put a few nuts on the rock, and hid himself, hoping to observe the birds' beaks and feeding behavior at close range. Although he waited three hours, the birds did not come back. ("It would be better to use caged birds," Peter noted tersely in his report.) But on Daphne the most famous beaks in the world were tapping him on the shoulder. He had Darwin's finches perching on his knee, studying him.

Peter was keenly aware that despite the birds' fame, despite their central place in the history of his field, no one had ever spent much time actually watching them. Darwin's insights were strictly retrospective. David Lack's were based mostly on inferences, museum specimens, and four months in the field. Bob Bowman had camped in the islands for less than a year.

"I think very quickly Peter saw the Galápagos as a gold mine," says Schluter. "Not just a wonderful place to be, but a gold mine, a treasure chest. Today, looking down the road at what one could argue is the most successful field study of evolution ever carried out, you ask yourself, Just how early in the game could Peter have foreseen all this, twenty years ago? I think maybe he had a glimmer of it from the beginning."

THAT FIRST YEAR the Grants and the Abbotts planned to stay in the islands for only a single season, so they worked fast, despite the heat. They studied twenty-one populations of Darwin's finches on seven islands. At each site, at dawn, they unfurled two or three mist nets. Mist nets look rather like badminton nets on bamboo poles, but of so gossamer a weave that they are almost invisible to birds. The team left the mist nets up throughout the cool of the morning. They furled them again when the island got so hot that the birds trapped and struggling in the nets were in danger of overheating. Most days it was that hot by eight o'clock in the morning.

The members of the team went to work on each finch they caught in much the same style they do today, armed with dividers, calipers, and a spring balance. No one had ever subjected Darwin's finches to so many different measurements and indignities, and no one had ever measured so many finches. Over the years, in fact, the Grant team's measurements of live Darwin's finches have far surpassed the number of specimens in the world's museums. Off the island of Isabela, for example, there is a group of four small islands known as Los Hermanos, The Brothers. In Los Hermanos alone, Trevor Price eventually measured twice as many living, breathing specimens of *fuliginosa* as repose today in museums. (The Grant team has also measured virtually every one of the thousands of museum specimens.)

In a study of variation, everything hinges on the accuracy of the measurements. Some subjects—the dome of the shell of a turtle, the webbed feet of a duck, the diaphanous gills of a fish—are hard to measure accurately. You measure once and measure twice, and your first number is quite different from the second. If the measurements are only good give or take a few percent, and if variations from individual to individual are much smaller than that, your study is doomed from the start.

Fortunately the finches and their beaks turned out to be not only easy to catch but also easy to measure. A long series of finch watchers could go back and measure the same bird, and they would all get numbers within a fraction of a percent of each other. Weight turned out to be unreliable, because a bird's weight goes up and down with the time of day and the time of year. But for other measurements the difference was almost always very small. For beak length, it was only a tenth of a percent.

"Hard facts" are those rare details in this confusing world that have been recorded so clearly and unambiguously that everyone can agree on them. The shape of a finch's beak is a hard fact.

The Finch Unit's measurements not only confirmed but heightened these birds' reputation for variability. They began to reveal how extraordinary Darwin's finches really are. The beak of the sparrow makes a useful comparison. Sparrows are closely related to Darwin's finches: some taxonomists place all sparrows and finches in the same family. One of Peter Grant's field assistants that first year was the Canadian biologist Jamie Smith. Ever since the early 1970s, Smith and his own team have been conducting a parallel watch, measuring song sparrow beaks on the tiny, remote island of Mandarte, British Columbia.

Smith has found that the beaks of song sparrows on Mandarte are all nearly the same length. It is rare to find a beak that is even 10 percent away from the mean. The probability of finding a sparrow that deviant is about four in ten thousand.

But in the Galápagos, the Finch Unit has discovered, the probability of finding a cactus finch with a beak 10 percent from the mean is much better than four in a thousand. It is four in a hundred. One of the Grants' world records in this respect is the depth of the upper mandible of the medium ground finch on Daphne Major. Here, the probability of finding a 10 percent deviation is one in three.

That is one of the most variable characters ever measured in a bird. And Darwin's finches are extraordinarily variable not only in the depth, length, and width of each mandible, and in the relative lengths of the upper and lower mandibles, but also in their wingspans, their body weights, and the lengths of their legs. Darwin's finches are even variable in the length of the hallux, or big toe.

Again, Darwin did not realize that these finches are so extraordinarily variable, because he did not collect enough of them to find out. Nor would Darwin have expected this result. He thought a small population would offer fewer variations from which nature can select. Hence he assumed that natural selection would be especially slow on a tiny oceanic island like Daphne Major.

So even during their first season, the Grants and Abbotts could see that Darwin's finches were more interesting than Darwin dreamed. And during that first field season, the island of Daphne Major sent the finch watchers an omen, a sign, a token of the difference a millimeter can make.

One day in April of that first year in the Galápagos, after a hard day measuring finch beaks, Ian Abbott climbed down to the welcome mat on Daphne Major to take in the view. The ledge is always encrusted with barnacles, each one a crude model of Daphne itself, a cone with a hole at the top. Because Abbott was sharing the ledge with these large, sharp barnacles, he wore a pair of old shoes. But because his wife and Peter Grant were the only other human beings on the island at that moment, Abbott wore nothing besides the shoes.

It was six o'clock in the evening, and the tide was coming in. Abbott squatted on his haunches, watching as the sun set on the neighboring island of Santa Cruz, and as hundreds of seabirds beat their way back to their roosts on Daphne Major. One millimeter beneath the future of Ian Abbott's genetic lineage, a single barnacle towered above all the rest. And as the first waves lapped the welcome mat, this great white barnacle opened its lid, extruded its feather duster, bumped into something, and nipped shut as powerfully as only a behemoth among barnacles can.

At least, this is how they tell it now on Daphne Major, where the story is still passed down from one generation of finch watchers to the next. They say Abbott screamed. Abbott bellowed. Abbott danced up and down on the welcome mat. At that moment he hated a barnacle as no man ever had before.

Chapter 4

Darwin's Beaks

What a trifling difference must often determine which shall
survive, and which perish!

— CHARLES DARWIN,
letter to Asa Gray

When Darwin was a student there, Christ's College was known as the
school of John Milton and the Reverend William Paley. Paley was required reading for a B.A., and Darwin read him over and over,
"charmed and convinced by the long line of argumentation." In fact,
Paley's *Evidences of Christianity* and *Natural Theology* gave Darwin "as
much delight as did Euclid."

*Natural Theology: Or Evidences of the Existence and Attributes of the
Deity Collected from the Appearances of Nature* had been a best seller
when it was first published in 1802. The book's first lines are sometimes quoted even today. "In crossing a heath, suppose I pitched my
foot against a *stone*, and were asked how the stone came to be there,"
Paley begins; "I might possibly answer, that, for anything I knew to
the contrary, it had lain there forever: nor would it perhaps be very
easy to show the absurdity of this answer. But suppose I had found a
watch upon the ground. . . ."

A watch would require more explanation than a stone. A watch,
says Paley, implies a watchmaker. Someone had to invent it; someone
had to put it together. And if that is true of a watch, Paley asks, how
much more so of the living things we find on the heath? Even the
simplest working parts of the smallest plants and animals go so far beyond our mortal powers of artifice that they imply "an artificer of ar-
tificers," a creator of creators, a God.

That was the world view of both Darwin and FitzRoy while they
were standing on the black lava of the Galápagos, where a live bird

looks almost as surreal as a watch on a heath. And afterward, that is how FitzRoy remembered the beaks of the Galápagos finches. "All the small birds that live on these lava-covered islands have short beaks, very thick at the base, like that of a bull-finch," FitzRoy writes in his *Beagle* memoirs (three volumes, with Darwin's memoir tacked on as a fourth). FitzRoy is mistaken, of course; his description fits only one out of the thirteen Galápagos finches, the heavy-duty lineman's pliers of *magnirostris*. But in any case, such a powerful beak—FitzRoy goes on to suggest—must be perfect for hunting and pecking on iron-hard lava and crushing berries for their juice. "This appears to be one of those admirable provisions of Infinite Wisdom by which each created thing is adapted to the place for which it was intended."

Darwinism is often spurned by the devout as a branch or prop of atheism. Yet Paley inspired Darwin at least as much as he inspired FitzRoy, and it was precisely this tradition of natural theology that led Darwin to the most original and unconventional step in his argument, his theory of the importance of variations.

If living things are well made, Darwin argued—if they are admirably adapted to their places in nature, contrivances more elaborate than watches—then even the slightest variations must make a difference to the individual animals and plants that are saddled with them. Some variations must help living things run better, some worse, and some—a very few variations, arising only once in thousands of generations—might help them fit into an entirely new spot in the economy of nature.

The beak of a bird makes a natural test of this step in Darwin's argument, not only because the beak is so easy to measure, but because it is so obviously vital to the life of the bird. Darwin's finches can't put food in their mouths with their wings. They can't use their claws either, any more than we can dine comfortably with our feet. They have to use their beaks. Beaks are to birds what hands are to us. They are the birds' chief tools for handling, managing, and manipulating the things of this world (*manus* meaning hand).

The shape of a bird's beak sets tight limits on what it can eat. Although the bones and the horny sheaths of the mandibles are a little more flexible than they look—a woodcock can poke deep down into the mud, part its beak at the very tip, and grab an earthworm—still, these are not many-jointed and articulate instruments like our hands. Each beak is a hand with a single permanent gesture. It is a general-purpose tool that can serve only a limited number of purposes. Woodpeckers have chisels. Egrets have spears. Darters have swords. Herons and bitterns have tongs. Hawks, falcons, and eagles have hooks. Curlews have pincers.

There are about nine thousand species of birds alive in the world today, and the variety of their beaks helped confirm Paley's belief in an inventive God. Flamingos' beaks have deep troughs and fine filters, through which the birds pump water and mud with their tongues. Kingfishers' beaks have such stout inner braces and struts that a few species can dig tunnels in riverbanks by sailing headlong into the earth, over and over again, like flying jackhammers. Some finch beaks are like carpentry shops. They come equipped with ridges inside the upper mandible, which serve as a sort of built-in vise and help the finch hold a seed in place while sawing it open with the lower mandible.

But plain or fancy, each beak can do only so much. The flamingo's beak is good for filtering pond water. The hawk's is good for ripping up a rabbit, a fox, or another bird. If the flamingo and the hawk ever tried to trade jobs, the hawk would drown in the pond scum, and the flamingo would get its eyes poked out.

Darwin extrapolates from big variations like these to the much smaller variations between individuals. According to his theory, even the slightest idiosyncrasies in the shape of an individual beak can sometimes make a difference in what that particular bird can eat. In this way the variation will matter to the bird its whole life—most of which, when it is not asleep, it spends eating. The shape of its particular beak will either help it live a little longer or cut its life a little shorter, so that, in Darwin's words, "the smallest grain in the balance, in the long run, must tell on which death shall fall, and which shall survive."

None of Darwin's readers doubted that the hooks, swords, spears, and pincers of the world's birds are of adaptive value. That was the pious, conventional, and commonsense view of Paley. But many readers did doubt that individual variations mean as much as Darwin says. He himself never actually saw a slight variation help or hurt the chances of an animal or plant in its struggle to survive.

Darwin does give one passage in the *Origin* the promising heading "*Illustrations of the Action of Natural Selection*." "In order to make it clear how, as I believe, natural selection acts," this passage begins, "I must beg permission to give one or two imaginary illustrations. Let us take the case of the wolf. . . ." Then he gives a few quick sketches to show that hypothetically, in a hard winter, when there is almost nothing else for a wolf to eat but deer, the fleetest and slimmest wolves would be expected to do best. Darwin makes similar arguments about nectar-bearing flowers and honeybees, all logical and hypothetical. That is the end of the section.

These sketches are so vivid that they enter straight into the minds of Darwin's readers. For years, those who accepted and those who rejected them felt no compelling need to go beyond them—and certainly not Darwin's bulldog, Huxley. "But the question now is:—Does selection take place in nature?" he asks, rhetorically, in one of his defenses of Darwinism. "Is there anything like the operation of man in exercising selective breeding, taking place in nature?" In answer he asks us to imagine what it must be like for a species of animal in nature, surrounded by fifty or one hundred others, with "multitudinous animals which prey upon it, and which are its direct opponents," and others preying on those, and still others indirect helpers, etc. He concludes that "it seems impossible that any variation which may arise in a species in nature should not tend in some way or other, either to be a little better or worse than the previous stock. . . ."

Even after half a century, Darwin's point about variations was still being defended with imaginary illustrations, and attacked for being imaginary. "The whole question [of the struggle for existence] has been discussed very largely from the *a priori* standpoint, throughout the whole period since the appearance of the *Origin of Species*," wrote the geneticist Raymond Pearl in 1911. "The 'rabbit with his legs a little longer,' the 'fox with the little keener sense of smell,' the 'bird of dull colors which harmonized with the background,' *et id genus omne*, have been made to do valiant service."

Darwin himself never tried to produce experimental confirmation of this particular point. It is at once extremely logical and extremely hard work to prove. Certainly he could not prove his case with his finches. He never learned much more about the details of their struggle for existence than he did during that first glimpse on San Cristóbal, when he watched the birds hopping together under the bushes, "scratching in the cindery soil with their powerful beaks and claws." If anything, those mixed flocks argued *against* his case. He saw finches with long thin beaks and short fat parrot-like beaks all hopping on the same lava, eating identical bird food. If beaks with such widely different shapes could handle and crack the same seeds, then what could it possibly matter if, among the parrot-beaked finches, one bird's beak was a little fatter than another's, or if, among the sharp-beaked finches, one was a little sharper?

On Daphne Major, for instance, the beak of the average *magnirostris* is 14, 15, and 16 millimeters in width, length, and depth. The beak of the average *fuliginosa* on Daphne is only about 7, 8, and 7 millimeters—that is, less than half as big. Yet Darwin saw both species eating the same food. If those two tools can do the same work, then what is the point of the Grants' measurements of two neighboring cactus finches, one with a beak 14.9, 8.8, and 8 millimeters, and the other 15.8, 9.7, and 9 millimeters? Variations that small would seem to mean nothing.

Natural selection is supposed to scrutinize the slightest variations in nature, "daily and hourly." But as far as Darwin could say after his five weeks in the Galápagos, natural selection is blind to the beak of the finch. No wonder he left them out of the *Origin*.

An ornithologist named Osbert Salvin looked over some museum specimens of Galápagos finches in the 1870s, four decades after Darwin collected them. It was Salvin who noticed how variable the finches can be, one from the next, in the length of their legs, in their wingspans and their weights, and especially in their beaks.

Salvin felt the Galápagos was "classic ground" (even in the 1870s). Having discovered that the islands' finches represent an extraordinary range of variations, he must have been disappointed that these variations seemed to have so little influence on the survival of the fittest. "The members of this genus," he wrote, "present a field where natural selection has acted with far less rigidity than is usually observable." Of course, the action of natural selection had never been observed at all.

Like Darwin, most of the series of scientific pilgrims who made their way to Darwin's islands arrived in the wet season, or what passes for a wet season in these desert islands. All of the bushes and trees were in leaf and flower, and there were plenty of seeds on the ground. The scientists watched closely, and they saw exactly what Darwin had seen. Most of the ground finches were hunting and pecking together beneath the half-naked bushes. All those different beaks were cracking the same birdseed.

One after another, this series of careful ornithologists concluded that the shape of the beak of a ground finch makes no detectable difference in the food it eats. "To look at the bills of these birds in the hand, we would conjecture wholly different diets," wrote the biologist and explorer William Beebe, who sailed out to Daphne Major through clouds of yellow butterflies in the wet season of 1923. "The small, delicate mandibles of *fuliginosa* would seem adapted to insect food, or at least small, rather soft seeds," he said. "At the other extreme, the huge beak of *magnirostris*, almost as large as the entire head, would be equal to the hardest of acorns." But both birds were eating identical foods. "What a mad country for birds and butterflies!"

It seems unbelievable now; but the action of natural selection is so easy to miss that Darwin's finches were considered by generations of ornithologists to be an exception, or even a counter-illustration, to Darwinism. In 1935, the hundredth anniversary of Darwin's visit to the islands, the ornithologist Percy R. Lowe gave a commemorative lecture before the British Association on the birds of the Galápagos. Lowe confined his talk to the Galápagos finches. He called them— apparently for the first time—Darwin's finches. But having read the by-now-extensive literature, Lowe declared his belief that the birds are not separate species at all, but "hybrid swarms." He thought their extraordinary variations would prove to be as meaningless as the varieties of coats in the stray mutts and cats in an alley. The beak of the finch offered "no scope for Natural Selection."

("Yes, he really said that, didn't he," Peter Grant says now. " 'No scope for natural selection.' Which was a wonderful way of stimulating people to go out and try to disprove him.")

Three years later, another British ornithologist anchored off San Cristóbal, Darwin's first island. Like Darwin he was a young man in his twenties. Lowe's lecture had piqued his interest, and although the Galápagos Islands had seemed "impossibly remote," he had been encouraged to make the trip by Julian Huxley, a grandson of Darwin's bulldog.

David Lack stayed through almost all of the wet season—one of the wettest wet seasons in the Galápagos in this century. The wet season is the finches' breeding season, and Lack saw a lot of breeding. He saw that the thirteen species of finches in the islands rarely interbreed. He even built aviaries in the long, hot, humid afternoons and tried to get the birds to hybridize inside them, but they would not cooperate. They seemed to be very particular in choosing their mates. So Lowe had been wrong to call them "hybrid swarms."

Still, even though the birds were not breeding together, most of the ground finches were feeding together, eating the same seeds. Lack had to agree with Lowe that the beak of the finch offers "no scope for natural selection." "In fact," Lack concluded, "there is no evidence whatever, in any of the island forms of Geospizinae [Darwin's ground finches], that their differences have adaptive significance." He wrote up this result in a monograph, but its publication was delayed by the outbreak of World War II.

It was some time after Lack got home to England that, like Darwin before him, he did a double-take. As he looked over his data he noticed that the species of finches whose beaks are most nearly identical do not live together on any of the islands in the archipelago. The cactus finch (*Geospiza scandens*) breeds on Daphne Major, for example, and also on all of the biggest islands in the archipelago except Fernandina; the large cactus finch (*Geospiza conirostris*) breeds on Genovesa and Española. Lack had never seen breeding colonies of both cactus finches on any one island. What is more, if two finch species with rather similar beaks do share an island, their beaks are more divergent in their measurements on that island than they are elsewhere. That is, the longer beak is longer than average, and the shorter beak is shorter than average, almost as if they were consciously trying to get out of each other's niches.

Lack found these patterns in case after case, not only in his own data books, but also in measurements of the thousands of museum specimens that had been collected since Darwin. Like Darwin, he could not see evolution in action, and he assumed it would be too slow to watch; but he could infer, looking back at the Galápagos, that something must be going on.

Lack was influenced in part by other biologists' studies of microcosms smaller than the Galápagos. Place two different species of *Paramecium* in a test tube, and come back in a few days. One species will have conquered the top of the test tube, and the other owns the bottom of the test tube, and the border in the middle is a no-cell's-land. Likewise with barnacles: one species takes the high tide line and another takes the low tide line.

Such experiments seemed to show that no two species eating the identical foods in identical ways can coexist peaceably in the same test tubes, on the same rocks, or on the same islands without one species driving the other to extinction. This is just the sort of competition and conflict that Darwin had imagined might lead to the extinctions of many branches and twigs on the tree of life; the branches in the middle would die off, and the survivors would bend and twist and diverge to either side as if to minimize competition by making themselves as different as possible.

Lack made charts of the beaks and their distributions in the archipelago. On island after island, Darwin's process seemed to him to have exterminated one or the other of two like-beaked species, or else to have pushed the survivors far enough apart to coexist. Whenever species with very similar beaks try to colonize the same island, Lack decided, they are thrown into competition. The struggle grows so bitter that one or the other species of finch is driven to extinction. But occasionally two like-beaked species evolve enough local differences that the intensity of their competition is reduced. Then both species survive.

Lack turned a classic negative case into a classic positive case, and helped to end the eclipse of Darwinism. In 1947 the very title of his monograph, *Darwin's Finches*, had a triumphant ring to it. Darwin's finches really are Darwin's. There is scope for natural selection in the beak of the finch.

The book had a powerful influence on specialists and on the general public, even though Lack had not actually seen natural selection in action any more than Darwin had.

DURING THEIR FIRST FIELD SEASON in 1973, the Grants and the Abbotts measured not only finch beaks but also finch behavior. They staked out eight sites of twenty-three thousand square meters. At each site they marked off a grid of reference points by tying red flagging tape to hundreds of cactus bushes and torchwood trees. Each morning they would crisscross one of the grids with binoculars, notebooks, and stopwatches, and see what the finches ate for breakfast.

The Grant team discovered that the ground finches were concentrating on about two dozen different species of seeds. So the members of the team put each of these two dozen kinds of seeds between the points of a vernier calipers and measured them as carefully as they measured the birds' beaks. They also measured the seeds' hardness with the McGill nutcracker. This is a gadget that Peter Grant designed with the help of an engineer at McGill University, in Montreal, his first teaching post. The McGill nutcracker is a pliers with a scale attached. Squeeze a seed with the pliers, and the scale shows how much force it takes to crack the seed open. Modern physicists measure force in a unit they named for the founding father of their science: the newton. To crack a grass seed, which is a speck about the size of a poppy seed, takes very little force, less than 10 newtons. A big cactus seed, the size of a peppercorn, takes more than 50 newtons. Cracking the toughest seeds in the Galápagos requires a force of 250 newtons, which is enough force to lift more than a thousand cactus finches into the air.

Peter Grant combined the measurements of seed size and seed hardness and rated each kind of birdseed as the finches might themselves, in a sort of Struggle Index. The small soft ones of *Portulaca* score the lowest on this index, only 0.35. The big hard seeds of *Cordia lutea* score highest, almost 14. Any of the finches can handle *Portulaca* in its beak, but very few are up to *Cordia*.

The Grant team also kept a census of the numbers of each kind of seed on the lava. To do this objectively they used a random-number table to select a single plot of lava, one meter square, somewhere in each grid. Then they counted every single fruit and seed they could find on that square of lava, whether it was dangling from the top of a cactus tree or lying in the middle of a cactus patch. Next they chose a much smaller plot within that square meter, again at random, and they sifted the hot cindery soil, collecting every fruit and every seed they found. Finally they withdrew to their tents and spread out their trophies on white trays to count one by one. And they repeated the whole routine fifty times.

"Most miserable piece of data we did," says Peter Boag, who, with his wife, Laurene Ratcliffe, joined the Grant group early on.

"Sift the dirt!" groans Ratcliffe. "Count every seed, every single seed! That's *Portulaca*. That's *Rynchosia*. That's *Setaria, Acalypha, Mentzelia, Heliotropium* . . . aaargh!"

"People think fieldwork is so romantic," Boag says, "but a lot of it is real slog. This was absolutely the worst."

They got to know the Galápagos birdseed so well that they could recognize the main species at a glance. They could often recognize a seed as it shattered in the tip of a finch's beak. "That's one advantage of the Galápagos," Boag says now. "You know exactly what the birds are getting. *That's* why we all want to work there. Not because it's nice. Because it's simple."

In most parts of the world, one might find two hundred species of plants in a single spadeful of earth. It would be impossible to find out exactly what a flock of birds is getting in its beaks as they move by the hour from lawns to woods to meadows to the banks of a stream. But on Daphne Major the finch watchers could get almost a God's-eye view of their flocks, which never went anywhere and never migrated for the winter. And of course when the watchers unrolled their mist nets, adds Ratcliffe, "every bird you caught was a finch."

"No one anywhere has duplicated the kind of fieldwork we did in the Galápagos—because it was so *simple*," Boag says. "Those ecosystems are stripped to the bare bones."

By the end of their first stay in the islands, the members of the Finch Unit thought they knew the finches' tastes in seeds, fruits, insects, leaves, buds, and flowers. On Daphne Major alone the team had watched and made notes while medium ground finches ate 4,000 meals. They knew exactly what the finches were eating, and they knew the size and shape of the beaks the finches were eating them with. And most of the ground finches were eating the same seeds and fruits, just as Darwin had seen in his first glance on San Cristóbal.

Before Peter Grant left the islands, the acting director of the Charles Darwin Research Station, Tjitte de Vries, gave him some advice. He reminded Peter that in the Galápagos the first half of each year is wet and the second half is dry. The Grant team, like Darwin, Beebe, Lack, and all the rest, had visited the birds in the wet season. But the dry season might be the time to watch life squeeze Darwin's finches.

Darwin's process can be hard for us to spot when nature is flush, when we behold "the contented face of a bright landscape or a tropical forest glowing with life," as Darwin writes in *Natural Selection*, ". . . & at such periods most of the inhabitants are probably living with no great danger hanging over them & often with a superabundance of food. Nevertheless the doctrine that all nature is at war is most true. The struggle very often falls on the egg & seed, or on the seedling, larva & young; but fall it must sometime in the life of each individual, or more commonly at intervals on successive generations & then with extreme severity."

THE GRANTS CAME BACK a few months later. Even from the air they could see the difference as they flew in to the small airport on the island of Baltra (built by the United States Sixth Air Force during World War II and maintained now by the Ecuadoran Air Force). Everywhere the lava was brown, black, red; almost no green below the highlands of Santa Cruz. At the research station, de Vries told them that there had been no rain at all at the station in the months of April, May, June, and July.

The Grant team recaptured many of the birds they had caught during their first trip, and dangled them again in the spring balance. The finches had lost weight, and when the members of the team counted the seeds on the same plots as before, they could see why the birds might be hungry. There was not much bird food on the ground any more. The plants had dropped all their leaves and seeds and had stopped making new ones, and the birds had eaten so many of the old seeds that their platter was almost clean. At the study site on the island of Genovesa, the volume of finch food was down by 84 percent.

Not only was there less food for the finches, there was also less variety. Only about half of their favorite kinds of food were left. And in the wet season most of the seeds on the ground had been so small and soft that the average seed had rated only 0.5 on the Struggle Index. The seeds that remained were mostly big and tough, and the average was higher than 6.

In the wet season every ground finch had the same favorite menu, seven kinds of soft seeds and fruits. Each ground finch had spent about half its foraging time on those seven. But now the ground finches spent only about a thirtieth of their time on them.

Magnirostris has the biggest beak and the most powerful jaw muscles of any of the finches. It is the only finch that is strong enough to break and rip off the Grants' metal bands from its ankles. *Magnirostris* was now concentrating on big, heavy seeds, the seeds that almost none of the other finches can crack.

The long, thin beak of the cactus finch is another of the most distinctive-looking beaks among the ground finches. Cactus finches were now taking advantage of their beaks' special talents and dining almost exclusively on cactus seeds.

It was the same with all six of the ground finches. Now that they were reduced to tough foods, the birds' tool-kit beaks were determining what the birds ate. They had become specialists, and each bird's specialty was set by the shape of its beak.

Smaller, local variations were making a difference too. The Grants can often tell which island a finch comes from just by glancing at it. *Fortis* on Daphne, for instance, is smaller than *fortis* on Santa Cruz, even though the islands are in plain sight of each other and within flying range for the finches. Meanwhile cactus finches on Daphne have narrower, finer beaks than cactus finches on Santa Cruz.

These variations from population to population are often much subtler than the variations between species, a matter of a millimeter or so. But they too make a mortal difference. They help determine what each population can do to get through the dry season. For example, the average length of a *fortis* beak on the island of Pinta is slightly deeper than on Daphne Major. Torchwood stones on the two islands are almost identical in size and hardness. But on Daphne, Grant says he has seen some of the *fortis* take as much as six minutes to crack open a single stone. That is a long time for a bird to struggle, and most of the time the bird just gives up after a while and drops the stone. On the island of Pinta, however, the *fortis*—with their slightly deeper beaks—are much faster with torchwood stones, and four out of five *fortis* can crack them. The difference in their beaks is only a millimeter.

Among finches of the same species on the same island, individuals often vary by smaller amounts than that. Now we are down to the level of variation that Darwin himself argued is the cornerstone of evolution. Not even Lack ever suggested that differences this small can matter in the beak of the finch. But Peter Boag made a clear and simple test of their significance a little later in the watch on Daphne Major, after hundreds of *fortis* had been banded. Boag walked around and around the island. Each time he spotted a *fortis* with a band on its ankle, he watched the bird until he saw it pick up a seed, and he wrote down what kind of seed it was. Boag found that in the dry season the birds with the biggest beaks eat the biggest seeds, the birds with medium-sized beaks eat medium-sized seeds, and the birds with the smallest beaks eat the smallest seeds: another Goldilocks-and-the-Three-Bears result.

ONE OF THE FINCHES' MOST BITTER STRUGGLES for existence is their running battle with a weed called caltrop. The Grants have made a case study of this. It is a classic demonstration of the war of nature, and in fact caltrop's name is rooted in the fields of war. For more than a thousand years, soldiers have sown battlefields with a certain kind of low-tech booby trap: spiky iron balls. Generally each ball has four spikes, so that one spike always sticks up to calk, or cut, a man's foot or a horse's leg. Roman charioteers threw caltrops behind them to prevent pursuit. Yankee pioneers sowed the grass outside their log cabins with smaller caltrops—some called them iron stars—when there were Indians around. Caltrop, the plant, also bears the Latin name of *Tribulus*, from the same Latin root as tribulation: *tribulare*, to afflict or oppress.

Like many other plants, including the star thistle and the water chestnut, caltrop defends its fruit with sharp spines. Each roundish fruit is divided into half a dozen sections, or mericarps, and as long as the fruit is still on the caltrop plant, each section holds the seeds inward, to the center, with the sharp spines facing out. When the fruit dries, these mericarps fall one by one to the ground. There is a single row of seeds nestled inside each mericarp, like peas in a pod. One mericarp holds as many as half a dozen large, nutritious, nutty-flavored kernels, each kernel in its own woody compartment, like chocolates wrapped in wooden foils inside a locked wooden box.

A caltrop's mericarp can be awkward in a finch's beak, almost as awkward as an iron star beneath a human foot or a horse's hoof. In

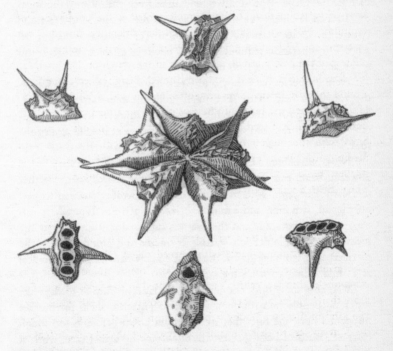

Caltrop. The large armored object in the center is a caltrop fruit. When it dries, it breaks into pieces called mericarps. The mericarps hold three to six seeds apiece. In this picture, finches have taken one seed from the mericarp at the bottom, and all of the seeds from the mericarps to its left and right, leaving small, black hollow cells where the seeds had been.
Drawing by Thalia Grant

fact, two species of finches on Daphne Major, the cactus finch and the small ground finch, have never been seen to try to open them. The only species that do attack mericarps are the large and the medium ground finches, *magnirostris* and *fortis*, and each species has its own tactics.

Magnirostris (whose beak is almost twice as wide and twice as deep as the beak of a *fortis*) picks up a mericarp, holds it near the midpoint of its beak, and squeezes its mandibles together. After a while the mericarp shatters into fragments. Then *magnirostris* picks up each fragment, holds it on one side of the beak, and crushes it. "When a *mag* is working on a *Tribulus*," Peter Grant says, "I can hear it cracking."

To crack a whole mericarp like this takes an average force of more than 200 newtons. Apparently that is more force than a *fortis* can muster. Instead it braces the mericarp against the ground and bites and twists the woody sheet that guards the row of seeds, as if peeling off a lid. This operation requires about 54 newtons of force, which seems to be about the best a *fortis* can do.

Neither species has it easy, and Grant has seen birds of both species use a rock to help them. The finch holds the mericarp in its beak, braces the upper mandible against a rock, and squeezes the lower mandible closed on the seed while pressing the upper mandible against the rock.

Now, *magnirostris*, because it crushes the whole mericarp in its powerful beak, can eat every last seed before it moves on to another mericarp. But *fortis*, with its smaller beak and weaker jaw, has to peel off the lid, exposing and eating one seed at a time. Typically it eats only one or two seeds and then moves on. It almost always eats the seeds in the same order too, starting at the narrow, pointed end of the mericarp and working toward the blunt end, as charmingly methodical as a child eating corn on the cob.

Watching these birds combing the dry lava for *Tribulus* seeds is like watching people hunt through a bowl of pistachio shells for the last unopened nuts, the ones that were thrown down before as too tough to crack. Birds of both species will often pick up a mericarp, work at it for a few seconds—sometimes longer—and then drop it and move on, like someone dropping a sealed pistachio nut back in the bowl. The finches prefer mericarps with only two spines, and mericarps with four spines are likely to be dropped. One indication that *magnirostris* has an easier time than *fortis* at eating caltrop, says Grant, is

Medium ground finch. From
Charles Darwin, *The Zoology of
the Voyage of H.M.S. Beagle.*
The Smithsonian Institution

that a *magnirostris* cracks many more mericarps than it rejects, while a *fortis* rejects many more than it cracks.

Darwinian competition is not only the clash of stag horns, the gore on the jaws of lions, nature red in tooth and claw. Competition can also be a silent race, side by side, for the last food on a desert island, where the competitors never fight one another, and the only sound of battle is the occasional crack of a *Tribulus* seed. Finches are locked in the most deadly competition even when they feed together in flocks. When times are hard, their lives depend on how efficiently they can forage for food—how little energy they can expend in getting how much energy in return. They are hungry, they are thirsty, and they are trying to keep their budget in balance. And as poor Mr. Micawber used to say, "Annual income twenty pounds, annual expenditure nineteen nineteen six, result happiness. Annual income twenty pounds, annual expenditure twenty pounds ought and six, result misery."

The race is to the swift, and *magnirostris* is clearly the winner. It can eat more than four seeds from two mericarps in less than a minute, while *fortis* gets only three seeds from two mericarps in more than a minute and a half. In fact *magnirostris* gets about two and a half times as much energy per minute, and because it gets more seeds out of each mericarp it has to hop around less too, which also saves energy.

Of course, *magnirostris* is bigger than *fortis*, not only in beak but in body size, so it needs more food. It needs one and a half times as much energy to meet its minimum daily metabolic requirements. But since the big beak of *magnirostris* earns the bird two and a half times as much energy, it still comes out ahead.

A few *fortis* have found a trick that helps them even the score. One of them sometimes trails a *magnirostris* around on the lava. As soon as the *magnirostris* cracks a mericarp, Grant says, the *fortis* rushes up, steals a piece, flies a little way off, and cracks it. Not every *fortis* on Daphne seems to know this trick; the Grant team has spotted only about half a dozen of them doing it. (Likewise on Daphne Major, cactus finches sometimes open cactus buds, and *fortis* has never been seen even trying to open them. But sometimes a *fortis* will wait beside a cactus finch, *scandens*, and after the *scandens* has torn open the bud, the *fortis* joins in.)

So the trials and tribulations of caltrop are not only harder on *fortis* than *magnirostris*; they are harder on some *fortis* than others. *Fortis* with bigger beaks can crack the mericarp and gouge out the seeds faster than those with smaller beaks. Tiny variations are everything. A *fortis* with a beak 11 millimeters long can crack caltrop; a *fortis* with a beak only 10.5 millimeters long will not even try.

"The smallest grain in the balance" can decide who shall live and who shall die. Between a beak big enough to crack caltrop and a beak that can't, the difference is only half a millimeter.

Incidentally, the beak of the finch may be exerting selection pressure on the caltrop itself. The Grants have not made a careful study of this. But out of curiosity Peter once compared the caltrop on the eastern rim of the crater, where there is heavy *fortis* traffic, and the caltrop on the northwest inner wall of the crater, about 20 meters down from the rim, where *fortis* rarely goes. Where there are many finches, each mericarp has fewer seeds, but it has longer and more numerous spines. In the steep, rugged, protected place, the mericarps have more seeds and fewer, shorter spines.

Peter suspects that the caltrop is evolving in response to the finches. Where the struggle for existence is fierce, the caltrop that is likeliest to succeed is the plant that puts more energy into spines and less into seeds; but in the safer, more secluded spot, the fittest plants are the ones that put more energy into making seeds and less energy into protecting them. The finches may be driving the evolution of caltrop while caltrop is driving the evolution of the finches.

According to the greatest authority on plants of the Galápagos, Duncan Porter, this species of caltrop comes from Africa. It may have traveled across the Pacific from island to island on the boots, pantaloons, and hairy legs of sailors, whalers, and buccaneers. If that is how it reached the Galápagos, then the very earliest date for its arrival on Daphne Major is 1535, which is the year the first European saw the Galápagos (it was the unlucky Fray Tomás de Berlanga, the third bishop of Panama, who was so glad to get away from the islands that afterward he did not even bother to name them).

We will never know when the first *Tribulus* seed reached Daphne Major. Probably it was not the year the good bishop collided with the islands, for he is unlikely to have seen Daphne Major, or to have stopped there if he did. So the war of beaks and spines on this islet may very well have evolved in the brief space of a few centuries— since the first seabird landed on the crater floor with a caltrop stuck to its webbed foot, or since the first human being sailed around the island a few times and planted a boot on the welcome mat.

"WHAT A TRIFLING DIFFERENCE must often determine which shall survive, and which perish!" Darwin wrote. To many of his critics this has seemed pure conjecture, but after a good part of a lifetime on Daphne Major, the Grants find it obvious. "I often think of piano playing," Rosemary Grant says. "I know I try to play the piano in spite of small hands, and how much easier it would be if my fingers were only a little longer. Or think of tweezers," she adds. Everyone in *El Grupo Grant* needs tweezers, because they collect cactus spines, especially during a nest census or a seed-and-fruit census, and some of Daphne Major's longest-suffering human inhabitants have come to consider tweezers "the most indispensable item of Daphne equipment." A complete tweezers kit includes a slant tip, a square tip, and a pointed tip. Often you can use any one of them to do the job of another, but that makes clumsy work and takes a long time. "And what a small difference in shape and size there is between the different kinds!" says Rosemary.

The Grants see the importance of variations on display not only on Daphne Major but throughout the archipelago. The second most closely watched island in their study is Genovesa, which the Grants began watching intensively in 1978. The work on Daphne is mostly Peter's, and the work on Genovesa is mostly Rosemary's. When they are alone the Grants call Daphne "Peter's Island" and Genovesa "Rosemary's Island."

On Genovesa, Rosemary has focused on the cactus finch *conirostris* (a finch that Darwin never saw). These cactus finches, like their congeners on Daphne, all eat more or less the same food when the food is cheap, but in times of famine they tend to specialize. Those with significantly longer beaks can hammer open the fruits of the cactus and probe the cactus flowers. Those with longer and deeper beaks can crack the big, tough cactus seeds. Those that have significantly deeper beaks than the others can strip the bark from the trees to get at the bugs beneath. It is one more demonstration that Darwin was right about the importance of slight variations. All this is precisely the kind of illustration that Darwin asks his readers to imagine in the *Origin*.

Once we accept that slight variations can help decide who lives and who dies, Darwin takes his thesis a step further. He argues that favorable variations will be more likely to be passed down. They will spread through the population, from one generation to the next, while

Two medium ground finches on Daphne Major: same species, same age, same island, yet one beak is strikingly deeper than the other.
Drawing by Thalia Grant

variations that hurt individuals in the population will tend to dwindle and die out.

During the eclipse of Darwinism this point seemed as much a matter of faith as the rest of Darwin's theory. Believers accepted it, skeptics rejected it. In the 1930s, for instance, the British evolutionists Robson and Richards analyzed the handful of studies that purported to show evolution in action. Robson and Richards concluded that even where natural selection might just possibly have been detected in action, the case studies did not prove Darwin's point because, in the opinion of Robson and Richards, the variations in question had not been passed down, and variations that are not passed down cannot lead to evolution.

Peter Boag has a background in genetics. After the Grant team had been watching Daphne a few seasons, Boag decided, as part of his thesis project, to try to measure the relation of parent bill size to offspring bill size in Darwin's finches. That is, he would measure how accurately the variations in their beaks are passed down—a factor that matters as much to their evolution as the presence of the variations themselves, or their influence on the lives of the individual birds. Improbable as it sounds, no one had ever actually tried to measure this factor, known in the jargon of genetics as heritability, in the wild. The more accurately the variations are reproduced—the more heritable they are—the faster the work of evolution could proceed among these finches. And without actually making the measurements, Peter Boag explains, "We didn't have any basis to judge."

Boag looked over all of the finch group's data, several years' worth, and he compared the sizes of the offspring and the sizes of their parents. He found that the body size of a finch does indeed depend very strongly on the size of its parents. A finch's size is highly heritable.

Boag also compared the birds' beaks to their parents'. The shape and size of the beak too is highly heritable. The beak of the finch is passed down faithfully from one generation to the next.

There was one loose end that could invalidate Boag's results. Suppose for the sake of argument that on Daphne Major in those years, *fortis* finches with bigger-than-average beaks were able to get more food. And suppose birds that eat better as babies grow up to be bigger-beaked adults. If so, a pair of big-beaked parents would have tended to provide more food for their babies, and their babies would have grown up big-beaked too. Big-beaked parents would have big-beaked offspring, small would have small, and yet the effect would have nothing to do with genetics. Despite the correlations that Boag had found, the size and shape of the beak of the finch might not be passed down from parent to chick.

This was nothing more than the age-old question of nature versus nurture, and Boag knew how to test it. If he took some eggs from a pair of big finches and put them in the nest of a pair of small finches, would the young grow up looking like their true parents or their foster parents?

Boag never did have a chance to perform this experiment during his watch on Daphne. And in retrospect, the finch watchers are glad he didn't, because their study is now so sensitive that the large number of egg switchings that Boag planned would have caused unnatural disruptions from that day to this. In fact, locally, Boag might have changed the course of evolution.

Instead, the egg switch was performed by Jamie Smith, after he left the Galápagos and set up shop among the sparrows on Mandarte Island. Smith switched many eggs from one sparrow nest to another, just as Boag had planned to do on Daphne. He found that the foster birds took after their true parents, not their adoptive parents. Being raised by a larger bird does not make you a larger bird. The young birds resemble their true parents, even though they are not raised by their true parents. This is very strong evidence that it is nature, not nurture, that plays the larger role in deciding the size of the sparrows and the shape of their beaks. As with the finches, the sparrows' beak variations get passed down with remarkable fidelity from one generation to the next.

Recent studies have shown that even the smallest details of bird life, everything from the exact size of the eggs to the number of the eggs and the date they are laid, are heritable too (at least to some degree). They are passed down from generation to generation in species after species of birds. This seems to be the rule rather than the exception in nature, just as Darwin imagined it to be, although not all variations in the living world are passed down as faithfully as the beak of the finch.

ORIGINALLY THE GRANTS HAD PLANNED to study Darwin's finches for a few months and lug home as much data as they could. Then they would try to sort out some of the forces that have made the birds what they are. In other words, the Grants were planning to take a snapshot. And if Darwin had been right about the slow pace of evolution, no one could ever take more than a snapshot. Watching these birds would be like watching the stars for an astronomer, or the mountains for a geologist. Even one hundred years in the Galápagos would be a snapshot.

But as the pieces fell into place, the Grants and their team began to understand that they had something worth watching. They would have to come back. The birds are exceptionally variable in their beaks. They are exceptionally sensitive to these variations. They pass on their variations with exceptional fidelity. Each of the requirements of Darwin's process, each of the prerequisites for evolution by natural selection, is heightened in Darwin's finches to an almost unnatural degree.

"Slow but sure moves the might of the gods," says the chorus in Euripides' *The Bacchae*. Slow but sure moves the power of natural selection, says Darwin. But here on Daphne Major, among Darwin's finches, the action might be swift and sure.

No one had ever stood watch before among Darwin's finches, so the Grants could only wait and see. As things turned out, they did not have long to wait.

Chapter 5

A Special Providence

... there's a special providence in the fall of a sparrow.

— WILLIAM SHAKESPEARE,
Hamlet

And for myself I am fully convinced that there does exist, in
Nature, means of Selection, always in action & of which the
perfection cannot be exaggerated.

— CHARLES DARWIN,
Natural Selection

Before he leaves Daphne's north rim, Peter Grant stoops and scans the
dirt by the path. With his floppy-brimmed sun hat and gray beard he
looks both cheerful and grave, reading the dust. He is hunting for a
Tribulus seed.

"There was a time when you'd just say, 'There's one, there's one,
there's one.' Now you've got to search for them," he says.

He kneels beside a *Tribulus* plant, or what is left of it. After almost
four years of drought, the plant has withdrawn to the roots. It looks
like a black claw hiding from the sun. All around it the lava is coated
with layers of old guano, and the glare on this white paint makes Peter
squint, even though the morning sky is still overcast. He brushes aside
a pebble or two.

"Here's a *Trib*," he says at last, holding it up in his palm. The *Tribulus*
plant dropped this mericarp during the last wet season, to wait for
the next wet season. Now the plant is a withered stump, and the mericarp, still waiting, has bleached to the color of driftwood.

Though it is guarded by two long sharp spines, this mericarp has
been chipped open at one end. Within the broken place, Peter can see

two dark pod holes, side by side, like tiny eye sockets, both empty. "Just two seeds taken out," he says.

At this moment, up and down the volcano, four hundred Darwin's finches are doing what Peter has just done. They are turning over pebbles, inspecting the lava, raking the cindery dust with their claws, sometimes poking their heads down into dark crevices, looking for the last bleached seeds. To open new ground, one of them will sometimes brace its head against a big rock and roll over another rock with its feet. A finch that weighed less than 30 grams was once seen rolling over a rock that weighed almost 400. That is like a man rolling a boulder that weighs one ton. It is the labor of Sisyphus, and unlike Sisyphus, Darwin's finches cannot keep it up forever. They are wearing down the horny sheaths of their beaks. Some of them have scraped their feathered crowns almost bald. Their occasional reward is a treasure like the object in Peter's palm, one more tough vitamin capsule, a husk with a few kernels no one else has eaten.

In the Grants' first four years on this island, they never saw the struggle for existence get this intense. Those were good years for Darwin's finches. By the end of the Grants' first season, for instance, there were about fifteen hundred *fortis* on Daphne Major. Nine out of ten of those *fortis* were still alive in December, just before the next rains came. There were also about three hundred cactus finches on the island that first April, and nineteen out of twenty of them survived the dry season and made it through to December.

Their fourth year, in 1976, was especially wet and green. There were great bouts of rain in January and February, and light showers in April and May, a total of 137 millimeters of rain, which is a good year for Darwin's finches.

The fifth year of the study, 1977, began well too. Rain fell right on schedule in the first week of January. Within days, green leaves unfolded and flower buds opened all over Daphne Major, with here and there a few caterpillars crawling on the buds, fast food for Darwin's finches. There were more than one thousand *fortis* and almost three hundred cactus finches on the island.

By this time the Grants' first pair of colleagues in the islands, Ian and Lynette Abbott, had gone back to Australia, and Jamie Smith had gone back to Canada. The Daphne watch had been taken over by Peter Boag and Laurene Ratcliffe. Boag was eager for the island's *fortis* to start laying eggs, because he needed more parents and offspring for his study. He had also received special permission from the staff of the National Park Service of Ecuador to perform his egg-switching experiment. After the *fortis* eggs hatched, he was planning to band the foster chicks. The next season he would measure their beaks' length, depth, width, and finish his Ph.D. with a bang.

After the first rain, a few pairs of cactus finches mated. (Cactus finches often breed before much rain has fallen, perhaps because they make most of their living from cactus.) The birds laid their eggs in nests they built in the cactus trees, and the eggs hatched fine.

Fortis do not breed until a little more rain has fallen. Peter and Laurene waited for the next good cloudburst, the one that would trigger the *fortis* to mate. But after the first week of January the sky above Daphne Major was like the sky that hangs over the Grants this morning, gray, low, and gloomily quiet. There was one more shower, a very light one, then nothing but clouds and heat.

The small rain that fell in the first week of the year did not settle into the soil. There is no place on Daphne's slopes for water to pool, and there is not much dirt to soak it up. The rain ran down the sides of the volcano as if pouring off a roof, and trickled away into the sea. Whatever was left was baked away by the sun or dried in the sea breezes on the rim and in the eddying winds that circle within the great bowl of the crater, stirred as if by a fire in the crater bowl each morning, as the rising sun begins to heat the lava.

Peter and Laurene checked on the nests when the cactus finch chicks were seven days old. Cactus finches build domed nests deep inside the cactus bushes, where they are well protected from owls. The chicks were peeping away with as much energy as they do every year. Peter and Laurene reached into each cactus bush, lifted the chicks out, set them down in a hat, and measured them one by one. In a normal year, flies and moths would have been flitting around the cactus trees by now, and the chicks' mothers and fathers would have been bringing bugs to the chicks. The air should have been so thick with bugs that three sweeps of a net through the air would come up with hundreds. But at the end of that January the island was still so dry that there were few flowers to bring out the insects. Three sweeps of the net came up with only a couple of bugs.

When Boag and Ratcliffe sampled the contents of the chicks' crops with an eyedropper they saw that most were almost empty. They found a little pollen, a few bits of flowers, or a seed kernel, sometimes a small spider.

All over Daphne, leaves shriveled, flowers wilted. Boag and

Ratcliffe had not planned to stay on the island this long. The way things usually went, they would go early in the year and band the fledglings in their nests on the seventh day. Then they would leave the island and come back later on to see how all the young birds were doing. But this year, Peter and Laurene were stuck. They could not leave the island until the *fortis* had their fledglings.

From the upper rim of the crater, Boag watched the horizon. The wind usually comes from the south, and it is blocked by the much larger and taller island of Santa Cruz, which lies 8 kilometers to the windward of Daphne Major. Storms drop most of their rain on the south side of Santa Cruz, so that the north side of the big island, and all of Daphne Major, lies in a rain shadow. Boag could see regular downpours along the coast of Santa Cruz. The island of Santiago, 30 kilometers to the northwest, where Darwin passed a fortnight, got soaked too that spring.

"We panted, drank water, and read books," says Laurene. "I read *The Agony and the Ecstasy.* Peter read every single book on World War II."

"That's the place to read *The Rise and Fall of the Third Reich*," says Peter.

A few scattered showers fell on a few scattered days, 24 millimeters of rain in all. This was not enough to move the *fortis* to pair off and mate, and it was not enough to fill the air with moths and flies. Two out of three of the cactus finches' chicks died in the nest, and those fledglings that did make it out stayed close to their parents for twice as long as usual, some of them for more than a month. They hopped beside their mothers and fathers and begged with piercing cries and much shaking of their wings.

During the previous June, when the island was wet and green, there had been more than 10 grams of seeds in an average square meter of lava. The finches had already eaten their way through many of those seeds during the dry season of 1976. Even if the rains had fallen as late as March or April of 1977 the seed supply would soon have rebounded, and the *fortis* would have begun to pair off and breed and lay eggs for Boag's experiment. But the weeks went by, and the rains did not come, and the birds did not pair off. Day after day they went on pecking over the same square meters for the same diminishing supply of seeds. By June of that year there were only 6 grams of seeds per square meter. By December there would be only 3 grams.

As they always do in dry times, the birds went on looking for the easiest seeds. But now they were sharing the last of the last of the pistachio nuts. They were down to the bottom of the bowl. In June of the previous year, four out of five seeds that a finch picked up were easy, scoring less than 1 on the Struggle Index. But as the small, soft, easy seeds of *Heliotropium* and other plants disappeared, the rating climbed and climbed, peaking above 6. The birds were forced to struggle with the big, tough seeds of the *Palo Santo*, and the cactus, and *Tribulus*, symbol of the struggle for existence, a seed sheathed in swords.

Back in 1973, it had been quite rare to see a *fortis* try to crack one of the iron stars of the *Tribulus*, and when one of the birds did try, it needed an average of almost fifteen seconds to crack the mericarp. Boag and Ratcliffe got out the stopwatch again. A *fortis* could now get the side of its beak across a corner of a mericarp and twist, twist, until the capsule chipped in less than six seconds. The birds were getting a lot of practice with *Tribulus*—that is, those birds that could handle it at all.

Some of the very smallest *fortis* on the island, the ones whose beaks were too small for caltrop, were poking around the *Chamaesyce* instead. The herb *Chamaesyce* has small soft seeds, but it also has a milky, sticky latex when its leaves are wounded and its stems are broken. These little *fortis*, along with the very smallest birds on the island, the immigrant *fuliginosa*, began hunting for seeds in the *Chamaesyce* in spite of its latex. The feathers on the crowns of their heads got so matted, gummy, and sticky that they rubbed off afterward as the birds raked the cinders and gravel looking for more seeds. Their bare scalps were exposed to the sun all day. Boag and Ratcliffe began to find little bald finches lying dead on the lava.

They kept up the routine of capturing and measuring finches, dangling them in the weighing cup and recording the numbers in their waterproof notebooks (that year they did not need them to be waterproof). By June, many of the birds' weights were down as much as a quarter from the June before. A large number of these finches had failed to molt, although by now they badly needed a new set of feathers. Some of their contour feathers were so worn that the down underneath was exposed, as Boag would report later in his now-famous paper on the drought of 1977. He and Laurene found dead *fortis* lying on the lava with feathers so disheveled they looked as if they had been combed the wrong way. The fraying of feathers was hardest on the smallest species, the *fuliginosa*, which when not eating the small, soft seeds of *Chamaesyce* were poking about for bugs in the lichen on the torchwood trees, Boag says, or hawking for bugs from a naked branch. They needed all their contour feathers for flight.

Even in good years, finch watchers have to be careful never to leave a bucket of water standing open in camp, or Darwin's finches will jump in and drown. On one island another Galápagos biologist, an iguana watcher, did leave a jerrycan open once, and in the morning it was full of finches. The bucket is like an oasis—it draws thirsty animals from far and wide. Once at the Charles Darwin Research Station a centipede a foot long crawled into an open bucket, and as grasshoppers hopped in, the centipede ate them one by one.

This year the whole camp on Daphne Major became an oasis of sorts. A flock of finches—mostly juveniles, birds born the green year before—hung around the finch watchers' tent and picked up crumbs. Peter and Laurene grew particularly attached to one female finch— "Number 1750 or something like that," Boag says. "She would follow us around the camp. She didn't make it through the drought, unfortunately."

Down on the crater floor the blue-footed boobies shifted their weight from one leg to the other to cool off their webbed feet. Boag stuck a thermometer into the ground, in the tortured shadow of a cactus, and the soil was hotter than 50° C (122° F). Even where there were seeds lying out in the open, the heat was keeping the finches from foraging there between the hours of 11 a.m. and 3 p.m. Meanwhile at night Peter and Laurene would be shivering in their tent if temperatures got down to 75° F, because their bodies were so used to the heat. Boag lay awake and wondered how his flocks were doing, just as biblical shepherds once did in a far-distant desert: "In the day the drought consumed me, and the frost by night."

Now and then a frigatebird harried a blue-footed booby out of its kill of fish. If the fish dropped on the island, as many as ten or twenty finches would flock around it. They also scavenged broken eggs and fresh booby guano. They hung close when the boobies fed their young and fought for the fish scraps, and when owls left something of their kill, finches fought over that too.

In other years the finches had ignored the lava lizards that scuttle about the rocks. But once that year, Peter and Laurene saw a female cactus finch eating a black lizard tail, and nearby they spotted a female lizard with a freshly broken stump. Some days later, they saw the same bird chase after another female lava lizard, pecking at its tail.

She might have started a new entry on the finch menu—but that was the end of the episode. Another time they saw a blue-footed booby wounded by a frigatebird. A *fortis* stood beneath the wounded booby and drank the blood as it dripped on a rock.

"We just sat there, month after month," Boag says. "At the time, we were depressed. We were losing the breeding season, so we wouldn't get a generation. Plus, all these birds were disappearing. We kept up doing the normal checks and censuses. But our feeling was not the thrill of seeing evolution in action, as one might conclude from reading the subsequent papers, but the moderate despair of doing a research project and seeing your birds *dying*."

All of the cactus finch fledglings died before they were three months old. Not a single *fortis* laid an egg or built a nest. Of course, this was the whole design of the study: to watch and see—"on spec," says Boag—if there were going to be any minor selective episodes. But now that a selective episode was definitely in progress, Boag was wretched. The egg switch would have been a wonderful experiment, and he was sure that watching natural selection in action would never make him Dr. Boag. At best, the events that he and Laurene were documenting this year would make a page of someone else's thesis a long way down the road. "We thought it very unlikely that we'd be able to measure it at all," Boag says. "We thought, watch ten years, and then *maybe*. So I didn't recognize what was going on—the magnitude of the effect."

THEY WENT HOME at last and went through their data. There were only five or six months between field excursions, so they had very little time before they had to go back into the field for another round: just enough to get the data out of their notebooks and type them up while they could still interpret their scrawls and scribbles. Peter and Laurene got their data out of the field books and saw who was left alive and who had died. But there wasn't much time.

Boag pondered the shambles of his thesis. To get a large enough statistical sample of family resemblances he had to measure many hundreds of adults and their young. Even in a normal year only a fraction of the adults that he measured would breed. Only a fraction of those that bred would build a nest that he and Laurene would find. He would band all of the nestlings he found, and about half of those would die. He might measure two thousand birds, and he might end up with only one hundred of their offspring. "They're very precious birds, if you see what I mean," he says now. "Each dead bird is a lost data point. So that was my main concern. I was not thinking of each bird as a data point for *selection*. I was thinking of it as a lost data point for *heritability*."

When he and Laurene went back in January 1978, they watched the bright yellow flowers of the cactus trees open up all over the island, as they do at the turn of each year. All of Darwin's finches converged on the blossoms, gorging themselves on the pollen and drinking the nectar.

Boag and Ratcliffe did the usual census. They found fewer than two hundred finches alive on the island. Just one finch in seven had made it through the drought. They measured these survivors, and they also measured the mummified carcasses of the dead finches they found lying on the lava, banded and unbanded. The island is large enough that the finches that disappear each year usually disappear without a trace. They never found the body of their camp finch or most of the others. But they collected all the numbers they could, and at the end of that season, they went home again and they typed them into the computer for analysis.

Today when they lecture on the selection event of 1977, Peter Grant and Peter Boag plot the effects of the drought in three curves. The curves start in March 1976, when the island of Daphne Major was still green and lush. They end in December 1977, when the cactus flowered and the worst of the drought was over.

All through the drought the total mass of seeds on the island went down, down, down. The average size and hardness of the remaining seeds went up, up, up. The total number of finches on the island fell with the food supply: 1,400 in March 1976, 1,300 in January 1977, fewer than 300 in December.

Next they take the finches species by species. At the start of 1977 there were about 1,200 *fortis* on Daphne. By the end of the year there were 180, a loss of 85 percent.

At the start of the year there were exactly 280 cactus finches on the island. By the end of it there were 110, a loss of 60 percent.

Of the smallest ground finches, *fuliginosa*, there were a dozen on the island at the start of 1977, and only one of them survived the year.

They also plot the age of the survivors. Many of the survivors were the oldest birds on the island, and had been banded by the Grant team in 1973. Not a single *fortis* was born that year on the island, and only a single one of the *fortis* that had been born the year before survived the drought. Only one of the young cactus finches born the year before survived. The drought practically wiped out the cohort that was born the year before that too. That generation became rare, and rarer with each passing year, like steel pennies minted in a war.

At last, Grant and Boag look at the beaks of the survivors. They know how variable the beaks are. They know how much the variations matter. They know how the plants were doing, what the weather was doing, how life on the island was squeezing the finches. They know all these figures with unprecedented precision, as well as the dimensions of the finches that made it through the drought and the finches that did not.

Among *fortis*, they already knew that the biggest birds with the deepest beaks had the best equipment for big tough seeds like *Tribulus*; and when they totted up the statistics, they saw that during the drought, when big tough seeds were all a bird could find, these big-bodied, big-beaked birds had come through the best. The surviving *fortis* were an average of 5 to 6 percent larger than the dead. The average *fortis* beak before the drought was 10.68 millimeters long and 9.42 deep. The average beak of the *fortis* that survived the drought was 11.07 millimeters long and 9.96 deep. Variations too small to see with the naked eye had helped make the difference between life and death. The mills of God grind exceeding small.

Not only had they seen natural selection in action. It was the most intense episode of natural selection ever documented in nature. One result was a bizarre tilt of the sex ratios on the island. At the start of the drought there were about 600 males and 600 females. By the end of the drought more than 150 of the males were still alive, but only a pitiful remnant of the females. Males are typically larger than females by 5 percent, with proportionately bigger beaks, so the males generally had an edge.

In other words, among the males the biggest survived, and among females the biggest survived, but many more males survived than females. And what made the difference between life and death was often "the slightest variation," an imperceptible difference in the size of the beak, just as Darwin's theory predicts.

Many people—even biologists, even today—find the power of slight variations hard to believe. "Once, just as I was beginning a lecture," says Peter Grant, "a biologist in the audience interrupted me:

'How much difference do you claim to see,' he asked me, 'between the beak of a finch that survives and the beak of a finch that dies?'

" 'One half of a millimeter, on average,' I told him.

" 'I don't believe it!' the man said. 'I don't believe a half of a millimeter really matters so much.'

" 'Well, that's the fact,' I said. 'Watch my data and then ask questions.' And he asked no questions."

"*None*," Rosemary agrees. "And he sat there scowling, fidgeting and talking the whole time."

NATURAL SELECTION BY ITSELF is not evolution. It is only a mechanism that, according to Darwin, can lead to evolution. As Peter and Rosemary Grant put it, natural selection takes place within a generation, but evolution takes place across generations.

In the drought of 1977, they had seen and documented natural selection in action. The decimation of the finches by selection had been as ruthless as the aristocratic breeder of bulldogs in Darwin's day who said, "I breed many, and hang many."

But the finch watchers did not yet know if the episode would translate into an evolutionary change. They only knew that, according to theory, it was possible, since the beak variations are heritable: the changes that are wrought upon one generation can be passed on to the next, becoming muffled and compressed or stretched and warped, over the years, as they pass down the line of the generations and onward into the future.

This is a step the founders of Darwinian theory considered logically inevitable, but which many of those who came after Darwin have doubted. Raymond Pearl again: "In the minds of an astonishingly large number of people, which number includes some rather great names in the world of science, it is precisely the same thing to show that something logically must be so, as it is to show that it is so. If the formal rules of logic are satisfied, truth seems to them to be thereby established. No further evidence is demanded. As everyone knows, this attitude led practically to the intellectual bankruptcy of the whole evolution theory. . . ."

On January 9, 1978, the clouds rained at last on Daphne Major. More than 50 millimeters fell that day. The rain fell on nothing green, only rocks and dead-looking trees and withered weeds. The rain streamed down the sides of the mountain.

All over the island, the male finches, the drought's survivors, did what they do each year in the first bout of rain. They flew to the highest points in their territories: to the crown of a tree that rises from a fissure toward the sky, or the crazy steeple of a cactus on the summit of a rockfall. Perched on these wet command posts, looking as skeletal and tattered as they had ever been in their lives, each cock opened his famous beak, like a rooster in a barnyard in the first light of day, and began singing in the rain.

The rain transformed the island. Within a week there were leaves and flowers on the torchwood. Green stems shot up before the finch watchers' eyes. "*Merremia* seedlings were 5 centimeters tall," Peter Boag reported afterward; "*Portulaca* was in complete leaf, and *Amaranthus* was 2 centimeters tall." Soon *Tribulus* and a dozen other plants had green fruits or seed heads, and the buds of the *Portulaca* were crawling with bugs. From the sea the dusty sides of the old volcano turned from its dull predawn shade of brown to a noon-green emerald, a tropical paradise.

Not a single pair of finches on the island had survived intact. The drought had taken the life of at least one member of every couple. But as the rains fell, many of the females' beaks began turning dark brown, and the males' beaks turned black, signs that the birds were ready to mate once again.

The males built nests in the cactus and sang for days on end from the highest cactus top they owned. The females hopped from territory to territory, inspecting the nests and presumably the singers.

Of course the skewed sex ratio put a spin on that breeding season. Among *fortis* there were now six males for every female. Each female could choose among many males, but only one male in every six could win a female.

The males flew after the females that visited their territories in what the finch watchers call the "sex chase." The females hopped around and flew around to visit nest after nest, and took part in chase after chase, before one by one they each settled down with a single male.

Again the finch watchers watched and measured. They found that the males the females had picked were not a random sample, any more than the ones the drought had spared were a random sample. The successful males tended to be the largest of the large. They were the males with the very blackest, most mature plumage and the ones with the deepest beaks.

Because of the crazy sex ratio, most of the males were left out; only a very small subsample of survivors had a chance to mate. But every single one of the eligible females was able to pair off. One female cactus finch set a record for the island, breeding five times and producing thirteen young.

Now it became of great significance that variations of body and beak are passed on from one generation to the next with fidelity. As a result, the males' unequal luck in love helped to perpetuate the effects of the drought. The male and female *fortis* that survived in 1978 were already significantly bigger birds than the average *fortis* had been before the drought. Of this group the males that became fathers were bigger than the rest. And the young birds that hatched and grew up that year turned out to be big too, and their beaks were deep. The average *fortis* beak of the new generation was 4 or 5 percent deeper than the beak of their ancestors before the drought.

In the drought of 1977 the Finch Unit had seen natural selection in action. Now in its aftermath they saw evolution in action, in the dimensions of the birds' beaks and in many other dimensions too.

After that, the watchers on Daphne Major had to keep watching. They had to keep coming back. Not only is Darwin's process in action among Darwin's finches, not only can natural selection lead to evolution among their flocks, but it leads there much more swiftly than Darwin supposed possible. The finch watchers had to find out what would happen next.

But even if they had quit there, what they had seen on Daphne Major from 1973 through 1978 would be enough to fill an old and rather embarrassing lacuna in Darwinism. In 1909, at the centenary of Darwin's birth, during a scientific meeting at Cambridge University, the German biologist Weismann asked whether natural selection can really explain the first small steps of evolutionary change. "To this question even one who, like myself, has been for many years a convinced adherent of the theory of selection, can only reply: '*We must assume so, but we cannot prove it in any case.*'"

Several human generations later, Darwin's descendants no longer have to assume. They can supply what Weismann wanted with case after case and, now, with Darwin's finches.

One of Peter Grant's graduate students, Trevor Price, has reviewed those early years using powerful mathematical tools that did not exist in the 1970s. These tools allow investigators to disentangle which among all the changing features of a bird or a fish or a fern is most strongly selected during a selective episode. That is, they help show investigators which changes in the living form were essential and which were simply following along, which parts of a living form were the targets of selection. This technique (it is known as partial regression analysis) was developed in 1983 by the evolutionary theorists Russ Lande and Steve Arnold. As soon as Lande and Arnold published the technique, Price applied it to Boag's drought. This reanalysis brought the evolutionary event into even sharper focus.

Price knew that the survivors and their offspring were larger in weight, wing length, tarsus length, and also in beak length, depth, and width. However, partial regression analysis shows that not all of those were selected by the drought with equal emphasis. During that terrible drought on Daphne Major, among *fortis*, nature was selecting most powerfully for bigger body size and deeper beaks. Nature was *not* selecting for longer beaks; a *fortis* with a long beak had no special advantage in the drought. And nature was rejecting the birds with wider beaks. So it was big birds with deep but relatively narrow beaks that were favored: perhaps, Peter Grant writes, "because a narrow yet deep bill was the best instrument for performing the difficult task of tearing, twisting, and biting the mericarps of *Tribulus* to expose the seeds."

So the birds were not simply magnified by the drought: they were reformed and revised. They were changed by their dead. Their beaks were carved by their losses.

In most places on this planet, the sight of a dead bird is so rare that it shocks us, even scares us. We recoil as if something has gone wrong in the cosmos, as if a shutter has creaked open that should have been kept closed, exposing a shadow world beyond our world, a place we were not meant to see.

But on the desert island of Daphne Major, dead birds are commonplace. They are everywhere. The lava is always littered with wishbones and beaked skulls. Whole seabirds lie outstretched here and there as if still in flight, odorless and mummified like feathered pharaohs in the dry and desiccating heat. Each generation lies where it falls, and the next generation builds on the ruins of the one before. They hatch in a morgue, breed in a crypt, and lie down with their ancestors, as if here not only life but death too is asking to be watched.

Evolution discloses a meaning in death, although the meaning is like some of the berries that Darwin tasted in the Galápagos, "acid & Austere." There is a special providence in the fall of a sparrow. Even drought bears fruit. Even death is a seed.

Chapter 6

Darwin's Forces

> And the more we learn of the nature of things, the more
> evident is it that what we call rest is only unperceived
> activity; that seeming peace is silent but strenuous battle. In
> every part, at every moment, the state of the cosmos is the
> expression of a transitory adjustment of contending forces; a
> scene of strife, in which all the combatants fall in turn.
> What is true of each part, is true of the whole.
>
> — THOMAS HENRY HUXLEY,
> *Evolution and Ethics*

The next watch fell to Trevor Price. He landed on the island early in 1979, and made camp in the craterlet on the east rim, the same small flat sandy spot where the Grants, and the Abbotts, and Peter Boag and Laurene Ratcliffe had camped before him.

Because he was standing watch after the others, Trevor could stand on their shoulders. He inherited all their painfully accumulated evidence demonstrating that variations in the finches' beaks are important, and that natural selection moves swiftly and surely among them. As Peter Grant's latest graduate student he also inherited the cooperation of the Charles Darwin Research Station, the routine of the *pangas* and *chimbuzos*, the names of local fishing captains, and (on buggy nights) the martial arts of the tent. Camping in the islands, he and his field assistant Spike Millington swatted mosquitoes beneath their tattered canvas with "Comparative Ecology of Galápagos Ground Finches," by Abbott, Abbott, and Grant, published in the *Ecological Monographs* of 1977, and already a classic.

After Trevor's first year on the island, seven finches in every ten were banded, and after the next year, nine out of ten. Trevor was the first who could recognize every single finch on the island on sight.

Ever since Trevor, the finch watchers have been able to recognize at a glance not only the banded birds but even the handful of rogues that are still at large. When a Finch Unit graph shows 1,250 finches on the island in a given month of a given year, it is not an estimate. The watchers have made a head count, like shepherds in a fold.

Trevor was able to follow Darwin's finches more closely than anyone else before him, not only because he had the benefit of so much data, and because so many of the finches on Daphne were banded, but also because after the drought of 1977 there were only a few hundred birds left. He could spot every one of their nests before its dome was woven. He had time to peer into it, mark its crook in the cactus tree with red flagging tape, and check it often on his rounds—around and around the desert island. Very few young fledglings hit the lava without one of Trevor's bright-colored bands around their ankles, the bands that would identify them as surely as rank, file, and serial numbers—or given, middle, and family names.

During Trevor's watch the flocks on Daphne Major became known so comprehensively and microscopically that the whole island seemed as small—for a brief interlude, at least—as a Petri dish. He moved toward the kind of near omniscience we expect in a laboratory. Now he could begin to see what happens to Darwin's finches when they are squeezed by not one but several conflicting selection pressures at once. For there is more than one force at work at a time in evolution, and their collisions are unruly.

Trevor measured all the finch chicks on Daphne when they were eight days old. He measured them again as eight-week-old juveniles, and as eight-month-old adults. He could see that the beak of the finch is full grown, or very nearly so, at eight weeks. If a young *fortis* has a beak depth of 9.45 millimeters at eight weeks, the bird will still have a beak of 9.45 millimeters at eight months, and at eight years, if it lives that long.

However, looking over his data, and the data of finch watchers before him, Trevor noticed something peculiar that the others had missed. When he looked at each generation of *fortis*, he saw that the average beak depth did not stand still as the birds grew up. In 1976, for instance, the average beak depth of the juvenile *fortis* on Daphne had been 9 millimeters. Six months later, the average beak depth of that same cohort of birds had dropped to 8.73 millimeters.

The individual birds had not changed. But the cohort as a whole had changed, because the smallest of the young birds, the ones with the shallowest beaks, were surviving, while the biggest, the ones with the deepest beaks, were dying. Trevor saw the same thing happen to generation after generation of young Darwin's finches. Not every small young finch survived, and not every big young finch died, but the small ones were the most likely to succeed.

After some thought, Trevor figured out why. At that tender age, the young finches' skulls and beaks are still soft, like the skulls and jaws of human babies; the bony puzzle pieces are not yet firmly sutured together. So even the biggest birds cannot crack the big hard seeds that some of them will learn to manage later on. During their first dry season, as the small seeds on the island grow scarce, and as the biggest of the adults on the island begin to eat the biggest seeds, the juveniles still have to hunt and peck only small soft seeds. Even the very biggest juveniles will not pick up a *Tribulus* seed.

Big birds need more food than young birds, and big juveniles need the most food of all, because they are still growing. But because they are young, their big soft beaks do not help them get it. For a finch, being young and big is all liability and no opportunity. After a while, as each dry season drags on, some of these big juveniles get so thin and sluggish that the finch watchers on Daphne Major can reach down and pick them up with their bare hands.

So bigger is not always better. While these birds are young, natural selection forces them in the direction of small size. When they get older, selection can force them in the direction of large size. Trevor measured these small conflicting waves of natural selection as they passed through each generation, like ripples bouncing back and forth across the face of a pond, pushing each cohort first one way and then the other.

LIFE IS NOT SIMPLE, even for a flock of birds on a desert island. Just staying alive from one life-stage to the next is a full-time job. And of course survival is only the first step. After the birds get a little older they also have to meet, mate, and raise families—while continuing to survive. Sex adds a whole new set of struggles to the struggle for existence, and the pressures of sexual selection sometimes conflict with the pressures of natural selection.

Darwin mentions sexual selection in the *Origin*, but he writes about it at greater length in *The Descent of Man*, published in 1871. In fact the subject fills half the book, whose complete title is *The Descent of Man, and Selection in Relation to Sex*.

In one way the process that Darwin called sexual selection is less harsh than natural selection. The worst penalty in the game of natural selection is death. The worst penalty in sexual selection is life without a mate. But then, a failure to mate is equivalent to genetic death.

In the dry season, natural selection metaphorically scrutinizes these birds, "daily and hourly," as they strive to keep body and beak together. Some birds make it, and some don't. In the wet season, which is also the breeding season, the survivors are scrutinized daily and hourly by one another, not metaphorically but literally, as males begin jousting for territory, building nests, and singing from the highest cactus in their territories, while females troop by and inspect the males' nests and plots of lava and listen to their songs.

In other words, as soon as nature stops selecting among these birds, the birds start selecting among one another. Again, some make it and some don't.

Darwin was convinced of the power of sexual selection, just as he was convinced of the power of natural selection. But he never saw evolution happen through either process, and his sexual selection theory went into a long eclipse after his death. It began to reemerge only after *The Descent of Man* was reprinted in a centennial edition in 1971.

The process has now been demonstrated in action many times and in many places, and one of the most dramatic demonstrations, of course, was the aftermath of Boag's drought on Daphne Major. There, because the females on the island chose only the largest males with the largest beaks, the process of sexual selection worked in the same direction as the process of natural selection and magnified it.

The skewed sex ratio on the island, a legacy of the drought, lasted a long time. During Trevor's watch on the island, the male *fortis* outnumbered the female *fortis* by two to one, or even three to one. This provided Trevor with an ongoing demonstration of the power of sexual selection. In fact it turned the island into as dramatic a theater for the comedy of sexual selection as the drought of 1977 had been for the tragedy of natural selection.

No females went without a mate in those years, and some of the females also had a second mate on the side. But males all over the island languished through the wet seasons without a mate, building nests in the cactus trees on their territory and singing from the highest point and winning nothing. Not one of the males on the island ever had two mates at a time, as far as Trevor could tell—and he was watching as closely as a Washington reporter.

Once again, Darwin's finches were not only meeting but surpassing Darwin's expectations. Darwin had assumed that the pressures of sexual selection would be higher in polygamous species than in monogamous species. Among the sea lions of the Galápagos, for instance, one male has a harem, and all the rest are bachelors and out in the cold—so the selection pressure among the males is enormous. Among more or less monogamous birds like these finches, on the other hand, Darwin assumed (reasonably enough) that the pressure of sexual selection would be milder, because there would usually be about equal numbers of males and females to pair off in each generation. Yet because of the remarkably skewed sex ratio after the drought the pressure of sexual selection on Daphne Major approached the intensity it attains among sea lions, where males play winner-take-all, losers-take-nothing. Year after year, many of the males remained bachelors, while other males mated and fathered many chicks.

Trevor could not tell, by eye, why one finch was winning round after round of sexual selection and others were losing. But after he left the island he entered into his computer the weight, wingspan, beak length, beak depth, and beak width of all the finches on the island in 1979, 1980, and 1981.

In two of those years, 1979 and 1981, the finches were breeding after having survived a terrible time, first the drought of 1977, and then the more moderate drought of 1980. During those breeding seasons, Trevor saw that it was the biggest males with the biggest beaks that were winning mates. The females were choosing the very features that had allowed the birds to pull through those hard times.

In one of the years of his study, Trevor also measured the sizes of the males' territories and their wealth—how many fruit- and seed-bearing trees there were in each male's territory. Here too he found a pattern. Males with bigger territories were more likely to win females than males with smaller territories.

Plumage mattered too. For a jet-black *fortis* male on Daphne Major, the chances of finding a female in 1979 and 1980 were better than fifty-fifty. But for a male that was not yet in full black feather—a male that still looked a little immature—the chances were less than one in three.

Black plumage indicates age and experience, and age and experience can make a difference in the number of offspring a couple can turn out. On the island of Genovesa, for instance, the nests of first-time fathers get dive-bombed by owls significantly more often than the nests of experienced breeders. Females with experienced, jet-black males are more likely to breed early in the wet season than females with first-time males. Experienced couples can sometimes lay not just one but two clutches of eggs before the breeding season is over.

If black males win more females, why don't all males turn black as fast as they can? Why do some of them turn black in their first year while others linger for years in plumage of such unattractive brown, in spite of the powerful sexual selection pressure that Trevor measured? The huge amount of variation in the finches' plumage suggests that there are hidden costs to wearing black plumage and hidden benefits to wearing brown.

While the females are flying around checking out each territory, the males are doing battle with all their neighbor males to establish and expand the boundaries of their piece of lava. Since a male in black gets into more fights than a male in brown, the most vigorous males are the black ones with a lot of land.

A male that stays brown a while may be able to avoid getting into so many fights, and set up its territory inconspicuously. "It does seem quite a problem for a male to establish a territory," Trevor says. "I once saw a young cactus finch wake up and give a few tunes at the corner of an old male's territory. The old male came for him like an arrow and hit him with his beak in full flight." Trevor saw one rather battered-looking male that held on to a small territory only as long as he was still in brown; as soon as he turned black, a neighboring male drove him away.

We think of the plumage of birds—the red of the male cardinal and the brown of the female, the green head of the male mallard and the brown of the female—as fixed and permanent, or as constant as anything else in the living world. To us they seem to stay the same as time goes by, like stones in a stream, which stand still hour by hour as the water flows. But while some of the birds' features are more or less set in stone, like the plan of their bodies—one beak, two wings, two legs—other features, permanent as they look, are the product of perpetually contending forces. They look solid, but they are as fluid as ripples on the stream. They are standing waves that grow or vanish, shift or disappear, with every change in the currents or the rocks.

*   *   *   *

THIS IS A STRUGGLE of struggles, or war of wars, that Darwin could only imagine, a war in which the forces of sexual selection wrestle with the forces of natural selection, pushing and pulling a living form this way and that, down through the generations. John Endler, the author of *Natural Selection in the Wild*, has been observing this conflict for years, in one of the most elegant and precise demonstrations of evolution in real time.

What the Grants are to Darwin's finches, Endler is to guppies. His guppies are not the variety they sell in pet stores (he considers those trash fish). His guppies live in northeastern South America, in the small streams that zigzag down the mountains of Venezuela, Margarita Island, Trinidad, and Tobago, flashing through steep, undisturbed green forests and then the broad spreads of the old cacao and coffee plantations, on their way to the Caribbean Sea and the Atlantic.

The male guppies wear black, red, blue, yellow, green, and iridescent spots in varying sizes, shapes, hues, and combinations. In fact their spots vary so much that they are like fingerprints: no two guppies are alike.

These spots, like the beaks of Darwin's finches, are heritable. Although the exact placement and arrangement of spots is unique, each guppy inherits its particular palette of colors, and also the general size and brightness of the ensemble, from its parents. The spots only show up on the males (they can be made to appear on the female guppies with testosterone treatments).

Like minute variations in the beaks of finches, the spots on a guppy are the sorts of details that one might imagine are beneath the notice of natural selection. Nature may scrutinize the slightest variation, but there are some things even Darwin's process cannot see. Design could not possibly govern a thing so small.

In the 1970s, while Peter and Rosemary Grant were watching the finches of the Galápagos, Endler began watching the guppies of Venezuela's Paria Peninsula, and Trinidad's Northern Range. There the streams run down the mountains roughly parallel, as if in a series of vertical stripes. The streams are clear, swift, and clean, deeply shaded by tropical evergreens and punctuated by waterfalls. Their beds are lined with brilliant, many-colored gravel, much like the floors of the fish tanks in the pet stores.

It is obvious to anyone who has ever tried to watch a school of these guppies against the parti-colored sands and pebbles of a streambed that the spots are excellent camouflage. In fact you could watch one of these clear streams for quite a while before you noticed the guppies at all, because they tend to swim close to the gravel while the sun is out.

The fish need this camouflage because they have seven enemies: six species of fish and one freshwater prawn. All seven of these enemies hunt guppies from dawn till dusk. The most dangerous is *Crenicichla alta*, a cichlid fish, which eats about three guppies an hour; the least dangerous is *Rivulus hartii*, which eats one guppy in about five hours.

Endler found guppies and at least a few of their enemies in almost every section of almost every stream, from the headwaters near the summit of each mountain to the plains and plantations below. Neither the guppies nor the guppy eaters can swim up a waterfall, and the population of each section of stream tends to stay put. (Sometimes a few fish get swept downstream, but none of them can get back up.)

High up near the headwaters of each stream, the only enemy the guppies have is the comparatively mild-mannered *Rivulus hartii*. But moving downstream, section by section, the population of guppies lives and dies in the company of more and more of its enemies, until down near the base of each mountain, the stream is loaded with all seven of the guppy eaters. So a graph of risk and danger runs with the current. For the guppies, the higher in the stream, the lower the risk; and the lower in the stream, the higher the risk. In stream after stream the intensity of natural selection is graduated in the same way: gentle pressure among the guppies at the top, violent pressure among the guppies at the bottom.

Endler saw that the streams would make a wonderful natural laboratory for the study of natural selection. He developed standardized methods of measuring guppy spots, as careful and ritualized as the Grants' methods with Darwin's finches. He learned to anesthetize and photograph each guppy he caught. (Like Darwin's finches, the guppies have met very few human beings, so they are easy to catch.) From the photographs he recorded the color and position of each spot of each and every male guppy, dividing each guppy into dozens of sectors to make a standardized guppy map that is easy to read, to tally, and to enter into a computer.

When Endler analyzed his surveys he discovered a pattern. The spots on each guppy look chaotic, but the spots of all of the populations of guppies in a stream, taken together, from the headwaters down to the base, have a kind of order. The spots on each population of guppies bear a simple relationship to the number of guppy eaters in their part of the stream. The more numerous the guppies' enemies, the smaller and fainter the guppies' spots. The fewer their enemies, the larger and brighter their spots.

The lucky guppies in the headwaters wear sporty coats of many colors, and each color is represented by big clownish splotches. Many of their spots are blue. These blue spots are iridescent, like the Day-Glo patches that cyclists wear; they flash as the fish swim, and they can be seen a great distance through the clear water.

Meanwhile the guppies downstream tend to wear conservative pin dots of black and red. The spots are almost vanishingly small. Most wear only a tiny amount of blue.

Endler looked at his data from stream after stream. In every one of them, the size and number of spots ran steeply downhill. And Endler drew the same sort of conclusion that Lack did when he noticed the patterns of beaks in the Galápagos. Endler thought he could see the hand of natural selection at work among the guppies. The greater the pressure from their predators, the more camouflage they wear; the less the pressure, the slighter the camouflage.

Of course, that interpretation did not explain why guppies are colorful at all. If they are in some danger everywhere, even in the headwaters, then why doesn't natural selection favor the best-camouflaged guppy everywhere?

The answer is that a male guppy has more to do in life than merely survive. It also has to mate. To survive it has to hide among the colored gravel at the bottom of its stream and among the other guppies of its school. But to mate it has to stand out from the gravel and stand out from the school. It has to elude the eyes of the cichlid or the prawn while catching the eyes of the female guppy.

The gaudier the male, the better his sex life. He is more popular among females, and he gets many chances to pass on his gaudy genes—as long as he lives. In a quiet spot near the headwaters of the stream his life is likely to be long and happy and he may father innumerable gaudy children. But in a spot near the base of the stream he may not father a single guppy before he vanishes down the gullet of a cichlid.

The quieter the colors of a male, the less luck he has in courting females. On the other hand he is likely to have more time to try, because the less he stands out among his own kind, the less he stands out among his enemies.

This is not just a problem for Trinidadian guppies. Wherever males court females, or females court males, whether the signals are a bright splash of color, as in guppies or red-winged blackbirds, or loud far-carrying songs, as in frogs and crickets, their broadcasts are always in danger of being intercepted by the enemy. Strong colors or loud calls can attract a mate from one side and a predator from the other. Every bullfrog calling in the night is in the dangerous spot of Romeo calling out beneath the balcony of the house of Capulet. A few species have found ways to finesse this problem. Among fish, some wrasses change color only very briefly, to flash a sexual signal in dangerous waters—the equivalent of a sexy whisper, *pssst!*

Looking at his guppy data, Endler read into them a struggle between two contending forces. Everywhere in the stream the gaudier fish produce gaudier young—pushing the next generation toward loud colors and self-advertisement. And everywhere in the stream the quieter fish produce quieter young, pushing the next generation toward modesty. In the relative safety of the headwaters the gaudier guppies live long enough to win many females before they are eaten, so the population evolves in the direction of greater and greater gaudiness, and almost every male wears a coat of many colors. But in the dangerous waters at the base of the mountain the gaudy guppies live such a short time that they are outreproduced by the modest guppies. So the whole population evolves in the direction of greater and greater drabness. Males court females at distances of 2 to 4 centimeters, and from there the little spots are visible; but from farther away the males blend into the gravel. So the small-spotted guppies can blend into the background in the eyes of their predators, Endler says, "yet still be visible and stimulating to females."

When Endler first began studying the guppy streams, he was in the same position as David Lack after the Galápagos. Endler could see patterns that strongly suggested the forces of selection at work. He did not actually see selection shaping the patterns, but the closer he looked the more he was sure that the hand that shaped the patterns really was the hand of natural selection. Within the broad patterns he kept finding curious subsidiary patterns. For instance in a few of the headwaters there are prawns. In these headwaters the guppies favor red spots. This red shift makes sense because although guppies and other fish see more or less the same colors that humans do, prawns and shrimp are red-blind—they cannot see the last band in our rainbows. So in those particular headwaters, male guppies with big red spots can show off to female guppies while hiding from the prawns.

Back in the 1940s, Lack made his selectionist argument about Darwin's finches without trying to measure it in the field to see if he was right. But Endler went the extra step: he decided to test the predictions of his theory by trying to detect these processes in action. He built ten ponds in a greenhouse at Princeton University. Four of the ponds were about as wide, deep, and long as the low-water territories of *Crenicichla alta*. The other six ponds were about the size of the headwater streams with the comparatively mild-mannered *Rivulus hartii*. Endler put black, white, green, blue, red, and yellow gravel in the bottom of his artificial ponds and pumped water through them to give them a current, like the streams in the wild.

Meanwhile, Endler collected guppies from up and down a dozen streams in Trinidad and Venezuela. In some places he took guppies that lived with just one predator, in some places guppies that lived with two predators, and so on up to the maximum, seven. He wanted stocks of wild guppies that had evolved under the whole spectrum of guppy menace, that were coping in the wild with every level of danger. He bred each stock in a separate aquarium.

When the artificial streams were ready for his guppies, Endler took five pairs at random from each stock and put them all together into two of his ponds to let them breed and mingle and get used to their new homes. Guppies can give birth at the age of five or six weeks, and a female guppy can spawn a lot of baby guppies, so it did not take long for the populations to double. After a month he took guppies from those two ponds and used them to stock two more ponds. A month after that, he had enough guppies to seed each of his ten ponds with two hundred fish per pond.

What he had done, in effect, was to shuffle and reshuffle the deck. He now had a highly heterogeneous assortment of guppies. They had all kinds of spots, and their spots were completely random with respect to the gravel at the bottom of their homes.

He let these guppies breed in their new streambeds for months. Then he added a few of their natural enemies to the streams, according to a careful plan. The evolutionary experiment had begun.

According to his prediction, the guppies should now evolve rapidly. The guppies in each tank should begin to look more like guppies that live with that same set of predators in the wild, they should come to look more like the gravel in their particular stream, and those in the most dangerous tanks should come to mimic the gravel more closely than those in the safer tanks.

After five months, Endler took his first census. He drained each stream, counting every male's spots and noting their position, anesthetizing them, photographing them, as he had done in the wild, and then starting up the stream again. Nine months later he took a second census. By that time nine or ten generations had passed in the lives of his guppies.

Some of the guppies were safe, with no enemies. These guppies got gaudier between the foundation of the colony and the first census, and they got gaudier still by the time of the second census. The males evolved more and more spots, bigger and bigger spots, wilder and wilder palettes of spots.

Meanwhile males in tanks with the dangerous cichlids evolved fewer and fewer, smaller and smaller spots. They were still visible to females, but they got less and less visible to cichlids, who strike from 20 to 40 centimeters away. These guppies mostly dropped the blue and the iridescent spots, their Day-Glo patches, just like guppies that live with cichlids in the wild. Endler measured these differences as meticulously as the Grants measure finch beaks. "Spot height, spot area, total area, and total spot area relative to body area also decreased significantly with increased predation intensity," he reports. The fish themselves changed size, too. Full-grown guppies in the dangerous tanks were smaller, while mature guppies in the safe waters were larger—again, just as in the wild.

Each tank had a different bottom: different mixes of gravel colors and different gravel sizes. In the pools with no predators the guppies did not change their spots to match the gravel—the opposite. Their spots evolved to be smaller than the big gravel and larger than the small gravel, making the males easier and easier to see, like chameleons in reverse. They carried more iridescent spots, and a wider palette of colors per fish, and generation after generation they looked less and less like their background, all of which is just what we would expect if they were competing for attention. Sexual selection was operating to make males as different from the gravel bottom as possible.

If only one force or the other had been operating, just natural selection or just sexual selection, the guppies would not have evolved in this remarkable way. Without natural selection all of the fish would have gotten gaudier. Without sexual selection none of them would have gotten gaudier. But the safe ones did get much more colorful, adding, in particular, blue spots. It is probably not a coincidence that guppies' retinas are exquisitely sensitive to blue. Almost all males carry some blue somewhere, even in the most dangerous waters—it may be a *sine qua non* of courtship.

The fish had evolved in Endler's greenhouse until they replicated the patterns that they display in nature, and they had done so in a very short time. Of course, Endler's streams were artificial. He had not seen natural selection in the wild. A skeptic could still argue that Endler was wrong about his explanation for the pattern in the wild. So Endler figured out a way to run the same sort of evolutionary experiment in nature.

Early in his fieldwork he had found a Trinidadian stream that contained the guppy eater *Rivulus hartii*, but no guppies. About 2 kilometers away was a second stream that contained both guppy eaters and guppies. Endler took a random sample of about two hundred guppies from one of the high-danger zones in the second river. He measured each and every one, as usual, and then he transferred them to the safe place in the first river. He took a sample of their descendants more than a year later, after a passage of fifteen generations.

The males in the safe stream were now much gaudier than their immediate ancestors, who were still living in the stream next door and coping with many enemies. The immigrant males wore bigger spots, and more of them, and each male sported a wider assortment of colors. Natural selection had acted just as predicted. Evolution had run as fast in the wild as in the greenhouse.

Everywhere in those streams, daily and hourly, natural selection in the form of cichlids and prawns is not just metaphorically but literally scrutinizing the male guppies. The result of enemy predations on each generation keeps pushing the males to blend in with the stream bottom. At the same time, daily and hourly, sexual selection in the form of female guppies is scrutinizing those same males. The result of their choices is that generation after generation of males is pushed to stand out.

Now it is clear why there is such virtually infinite range of variation in the way each individual male guppy is spotted. Many different random patterns of splotches will be equally good camouflage, because the streambed patterns are random too. It would not help the guppies to sport the same pattern as all the others, and in fact it would hurt them. If the guppy males all looked alike, their enemies could develop a search image—an inner template. They would search for that pattern, as we search for the face of a friend in a crowd. The rare misfit would have a great advantage. Meanwhile the females would go for the unusual males, too, and that would drive more and more diversity of patterns. So in this respect natural selection and sexual selection cease to oppose each other and push in the same direction: toward almost infinite diversity.

We see here an example of what Darwin saw in the wide world. He understood that his simple process can lead to the most bewildering and chaotic-looking diversification and variety—but underneath, the driver is as simple and plain and commonsense as ever, "small consequences of one general law leading to the advancement of all organic beings,—namely, multiply, vary, let the strongest live and the weakest die." The guppy experiments suggested to Endler what Darwin's finches were suggesting at about the same time to the Galápagos finch watchers: that natural selection can be swift and sure. The process is flowing along, all around us, much faster than Darwin ever dreamed.

Endler's study is leading him into deeper and deeper waters. He now suspects that the guppies' spots, their mating habits, and their color vision are all evolving simultaneously, with change in any one of these factors driving change in all the others. To measure variations in the guppies' retinas, Endler is collaborating with physiologists. These "hard science" types often remind him how "soft" the science of evolution is perceived to be by the outside world—even by biologists. "I was talking with someone in vision physiology the other day," Endler says, "and he told me, 'Wow, I had no idea that the subject was so *rigorous*. I had no idea that you actually did *experiments*.'

"We have a serious public-relations problem," Endler says. "People don't realize this is real science."

AS TREVOR PRICE'S WATCH DREW to a close, the members of the Finch Unit had logged almost a decade on Daphne Major. When they looked back, they saw this: in 1977, evolution by natural selection had made the birds bigger. In the wet season of 1978, evolution by sexual selection made them even bigger. Then one or the other or both pressures acting together made the birds on the island bigger yet in '79, '80, '81, and '82. The sum total of all the varied pressures of life on Daphne Major seemed to be pushing the finches in a single direction. At this rate, the birds would go on shooting into the future like a searchlight, growing larger and larger.

So the results of the finch watch presented a paradox. If there is strong selection for large size, year after year, why don't all the small finches turn into large ones? Why is there a small ground finch, a medium ground finch, and a large ground finch? Why doesn't every small become a medium and every medium a large? Or were they doing just that—had the course of evolution taken off in a new direction on this island just as the Grants and their students arrived to watch?

It did not seem likely that the trend would keep going like this. Surely the Grants had not begun to watch just at the moment that Darwin's finches were in the middle of a radical transformation. That would have been like training a new telescope on a distant star and watching it explode into a supernova before your eyes.

Something had to happen soon. The trend had to break. Some force had to whirl down upon the birds and force them back in the other direction.

Trevor Price was positive that with the next heavy wet season he would see something new. In the next solid rain he would detect the pull to oppose the push, the event that would reverse the trend he had seen in the run of dry years.

In 1980 Trevor had two bouts of rain. But that was not enough. It did not trigger much breeding. He needed just one good breeding bout.

Price got as desperate for rain as Boag had been on the watch before him. He paced the island for months on end, waiting for rain. Then he'd come into town—the fishing village of Puerto Ayora on the island of Santa Cruz—half-crazy because it hadn't rained. He'd be ragtag and barefoot, wearing an old striped shirt, checked pants, a beard that had never known a comb. He had a wild and friendly manner. He spoke atrocious Spanish with an impossibly strong British accent. He would hang out in town for months, making friends and having a good time. Everyone in Puerto Ayora knew him.

But he suffered through each drought convinced in his bones that in one good bout of rain he could make the discovery of his career. The finch watchers posted on the other islands of the archipelago were all rooting for him.

"That's the difference between field biology and physics," says one of Trevor's friends, philosophically. "You can be stuck waiting for rain.

You've got a beautifully planned research program, and all you need is the rain."

During Trevor's last year, 1981, he brought a new field assistant to Daphne Major, a Turkish mathematician named Ayse Unal. She put up with him mooning around for months waiting for the rain. And when rain came at last, in March, she watched him dance in it for two solid hours praising the heavens, a dervish of tangled hair and beard and ripped clothes, hollering into the rain.

But even then it was too little too late. Not many finches bred. Trevor still thinks about that last rain in the small hours, the iguana hours, the hours of what-might-have-been. If only that bout of rain had come in February, or January. If only his watch had lasted just one more year. "It would have been an amazing thesis then." ("It was an amazing thesis anyway," says Peter Grant.)

But it was time for another changing of the guard. Lisle Gibbs, a Canadian graduate student of Peter Grant's, took the next watch. He went down in December 1981 with a field assistant, to be sure to be there when it started to rain. They sat there, two men on a desert island, and waited half a year for the gray clouds to shower on them. Some rain fell—but not enough. "It was like waiting for Godot," Gibbs says now.

They went home to Michigan. And late in 1982, as Christmas approached, Lisle got a postcard from the village of Puerto Ayora. "It's raining."

Chapter 7

Twenty-five Thousand Darwins

... and the windows of heaven were opened. And the rain
was upon the earth forty days and forty nights.

—Genesis 7: 11–12

Lisle Gibbs no longer remembers how he got to the island. He simply
said to himself, "I'm going to be there." Seventy-two hours later he
leaped from a *panga* and landed on Daphne's welcome mat.

He had spent six months waiting for rain, and poor Price had languished three years. And in the end the rain had arrived so far ahead
of schedule that not a single human being was there on Daphne to see
it fall.

Lisle and a field assistant scrambled up the cliff and looked around.
The female finches' beaks were dark already, and the males' beaks were
black. The cactus trees were loaded with freshly woven domed nests,
and the nests were crowded with cheeping baby finches, the latest
generation of Darwin's finches: conceived and hatched before the
coming of the new year. "We were banding birds as soon as we
landed," Gibbs says.

No one in the Finch Unit had ever known a wet season to start
so early. And for the next few weeks the rains, though so long awaited,
were almost appalling. In retrospect it is as if the cup of the year to
come was so overfull that it was spilling backward. December 1982
drenched Daphne Major as never before in human memory—which
of course was less than a decade long. On the island of Santa Cruz,
next door, it was the wettest year's end since the founding of the
Charles Darwin Research Station in 1960: almost four times more rain

than ever before. Far to the north, on the island of Genovesa, when another team of finch watchers arrived on the very last day of the year, they found Rosemary's rain gauge already filled to overflowing.

Stupendous and stupefying thunderstorms rolled over the islands, terrifying the cattle on the island of Santa Cruz, for Galápagos clouds are mostly mute. There were landslides on the steep slopes of the islands' volcanoes, flash floods and overnight waterfalls. Cactus and *Croton* trees as old as the century were swept down the sides of the volcanoes, tumbling like twigs. Just above Gibbs's tent on the rim of Daphne Major the skies were low, black, loud, and flickering, and just beneath them the seas were high, and hurled green and white breakers up the cliffs. The island rode into the new year like a ship into a storm.

OF ALL THE CHRONIC PATHOLOGIES of weather on this planet the worst single repeat offender is El Niño, the Child, so called for its tendency to visit the Pacific shoreline of South America around Christmastime. During an El Niño a patch of abnormally warm water appears in the eastern Pacific and spreads until much of the eastern half of the ocean is running a fever of several degrees centigrade. Such a vast acreage of abnormally warm water stirs strange winds and weather virtually around the world; the Galápagos Islands sit in the epicenter, as it were, of the fevered water.

El Niños are born at irregular intervals, usually three to six years apart. The Grants had started their study just after the last Child left. And by January 1983 it was clear to everyone who knew the islands that the Child was back, and that this was to be no ordinary Child. The seas were hotter than normal even for an El Niño, and the clouds that rose from the Pacific seemed to the naturalist and sailor Godfrey Merlen, who lives in the islands, like "storms of the ocean itself." "The effect," he wrote in a memoir of that apocalyptic year, "was of huge mushroom-like trees growing from the sea with heights reaching many thousands of feet."

Lisle Gibbs not only got rain. He not only got an El Niño. He got the strongest El Niño in living memory—probably the strongest of the twentieth century.

On Daphne Major, his erstwhile desert island, Lisle waded through sheets of rushing fresh water. He slogged through black muck, frantically banding the new finch fledglings, which each day appeared faster than he and his assistant could keep up with them. They worked bareskinned and bareheaded through the torrents. Ponchos would have been useless, slick with rain on the outside and slick with sweat on the inside.

Unlike Boag and Ratcliffe in the drought, Gibbs in the flood was sure that an extraordinary selection event must be in progress, although he did not know what the finches would do. He banded fledglings in the rain and wondered.

"It was ridiculous," Trevor Price said long afterward, ruefully, looking at the pictures in one of Lisle Gibbs's papers. "It was like going from desert to jungle." Gibbs had vines growing up the tent poles, and he could see them grow from morning to noon and from noon to evening, a few centimeters per day. The *Croton* trees flowered not just once or twice but as many as seven times, so that each tree and bush set seven crops of seeds, and every one a bumper crop. One *Croton* seed fell to the ground in December, and by May the plant was level with the eyes of a tall man, whereupon it burst into flower too. By June the total mass of seeds on the island was almost a dozen times greater than it was the year before. It was as if nature had set out a dozen dishes for every finch, whereas in most years it sets just one. There were also more than five times as many caterpillars to eat, and every one of them about four times normal size.

It was too wet for cactus, and thickets of creeping vines smothered the *Tribulus*. So the big-seed crop crashed while the small-seed crop boomed. This was rags to riches for Darwin's ground finches.

"The birds went crazy," Gibbs says. "The year before there had been no breeding at all. Now they bred like hell." On Daphne, females produced up to forty eggs and fledged twenty-five young. The most prolific pair on Genovesa laid twenty-nine eggs in seven clutches, and twenty fledglings hopped out of the nest, a record for the island. In the steamy rains more and more of the birds were turning bigamous or polygamous. On Genovesa one female finch went through four males, one after the other.

The longer the copulatory frenzy lasted, the more finch fledglings were hopping on the wet lava for Gibbs to catch and band. By June there were more than two thousand finches on Daphne Major.

Most finches do not breed until they are two years old, and by then the finch watchers have gotten personally acquainted with each one of them. But in the middle of that breeding season, Gibbs and his assistant started to see banded birds they couldn't recognize. "Finally, we realized that they were kids—three months old," Gibbs says. The young birds they had banded in their first weeks on the island were pairing off and mating in the cactus bushes. No one on earth had ever reported anything like this: passerine birds are not supposed to breed in the same season they are born. But as the rains kept falling, almost every finch on the island was caught up in a breeding rush, like a gold rush. Some of the very young males staked out territories without a single cactus bush and still found mates. Many of the new pairs also successfully fledged young, especially young females who paired with older males. The youngest to breed was a female *fortis* less than three months old. She laid four eggs in her first clutch, and two chicks survived to leave the nest.

The cactus finches produced eight times as many clutches as they had in the previous breeding season. In fact, in that single year they produced more than half the young they would ever produce in their lifetimes. The numbers of cactus finches and *fortis* on Daphne Major rose by more than 400 percent.

It was a phenomenal year for finches all over the archipelago, from Wolf to Santiago, from Isabela to Española. On Genovesa, however, the gold rush had a dark side. The record numbers of eggs led to record numbers of deaths. Finches would sit in their nests in chilly showers for just so long; then they abandoned their eggs and sometimes even their begging young. Heavy rains and heavy winds snapped the branches, and down fell the babies, cradle and all. Meanwhile, in a development that proved tragic for the finches, many of the Genovesa mockingbirds took sick in the rain. Their legs and claws blistered and swelled with what was probably pox.

Most finches on Genovesa did not catch the pox, but even so they were hurt by the plague among the mockingbirds. Normally mockingbirds are bound together in tight little social groups, and the young unattached males help the older ones tend the nests. (The discovery that mockingbirds cooperate like this was made by the Grants' daughter Nicola on Genovesa when she was twelve years old.) But this year, as the plague felled the elders in group after group, the survivors drifted away. Eventually the adult mockingbirds settled down and built nests elsewhere. But the young kept on roving through the wind and the rain. Bob Curry, a mockingbird expert and a Ph.D. student of Peter Grant's, saw young mockingbirds hanging around the nests of the finches, sometimes singly and sometimes in little bands, like juvenile delinquents, "a steady stream of wandering mockingbirds." They ter-

A Galápagos mockingbird,
Mimus parvulus. From
Charles Darwin, *The*
Zoology of the Voyage of
H.M.S. Beagle.
The Smithsonian Institution

rorized finch nests across wide swaths of the lava. They drove off the
finch parents and ate the finch nestlings. One hapless cactus finch laid
eight clutches, a total of two dozen eggs, and not a single one of her
chicks lived long enough to hop out of the nest.

AFTER THE RAINS STOPPED, the islands began putting themselves back together. The sun beat down. Red mud caked and crazed.
By autumn the freshwater ponds in the uplands, where pintail ducks
had made themselves at home, were already drying up, and the innumerable rain pools in the islands' lava hollows, where shrimp had appeared as if by magic, were bone-dry again. The giant tortoises helped
return the upland landscapes to normal by crashing and smashing
down the weeds; they filled in the gullies by knocking down their
sides. As early as September a naturalist walking through the uplands
of Santa Cruz wondered, "Has this year really been?"

The numbers of finches on the islands were astonishing, and for
the first year or two after the El Niño there were still small seeds all
over the island to support their population boom, riches like the jewel

heaps in the caves of the *Arabian Nights*. But the skies that gave now took away. Only 53 millimeters of rain fell the year after the flood, and only 4 millimeters fell the year after that. The plants did not set half enough seed to replace the bumper crop from the year of the Child.

The finches had overshot the carrying capacity of their desert islands, and now Lisle Gibbs watched their populations crash. He went on observing the huge flocks of Darwin's finches on Daphne Major in 1983 and 1984, banding the newcomers and marking the deaths in his field notebooks with little crosses. Finches were dying right and left, as they had died in Boag's drought.

Would evolution continue to shoot like an arrow in the same direction, or reverse? How were the birds evolving now? He could not tell until he had accumulated a long enough record and fed it into a computer.

In September 1985 Lisle was back home at the University of Michigan, Ann Arbor, where the Grants were teaching in those days. It had taken Lisle a year just to enter all of the data from his waterproof notebooks into the computer. He had spent months and months checking and double-checking the data for errors and checking and rechecking the program with which he would analyze the data for evolutionary trends.

Now he ran the program. "I cranked out the numbers, and I was praying," he says. "I remember the actual moment when I hit the return key. After all that work . . ."

What he saw on the screen was so dramatic that at first he refused to believe it. He checked and rechecked. It was true. Natural selection had swung around against the birds from the other side. Big birds with big beaks were dying. Small birds with small beaks were flourishing. Selection had flipped.

Both big males and big females were dying, he noticed, but many more males than females—again, the reverse of the drought. Everything the drought had preferred in size large—weight, wingspan, tarsus length, bill length, bill depth, and bill width—the aftermath of the flood favored in size small.

At first, Lisle Gibbs and the Grants were not sure why the flood year dragged the birds backward like that, although it did make intuitive sense that an epic flood would undo the work of an epic drought. But eventually they came to understand why the flood favored small finches over big ones. With ten times more small seeds lying around, the large finches had trouble finding large seeds. They could still eat small seeds, of course, but they had the tools for large seeds, and they had a lifetime of experience hunting and cracking large seeds; and of course being big birds they had to eat many more small seeds to stay alive.

So as seed supplies ran lower and lower, the bigger birds had more and more trouble. They were in the same sort of predicament that big young finches experience in their first few months of life. They paid dearly for their large size, because it gave them a larger appetite, and they could not make it up to themselves with their large beaks. Some of the large-beaked birds made the shift, but slowly and not as well as those with the right equipment.

The net result of natural selection during Gibbs's watch was as stark as during Boag's drought. The birds took a giant step backward, after their giant step forward.

A terrible drought like the one in 1977 may come once or twice in a finch's lifetime, and an El Niño like the one that came in 1983 is a once-in-a-lifetime event. So having witnessed both the year of the drought and the year of the flood the finch watchers were now staring at an extraordinary picture. Clearly, selection pressures on a creature in the wild are far more intense in some years than others. But more than that, even the most intense selection pressures can actually reverse

A Galápagos iguana. From Charles Darwin, *The Zoology of the Voyage of H.M.S. Beagle.*   The Smithsonian Institution

themselves during the creature's lifetime. Not only can evolution push a species fast in one direction. Evolution can reverse direction and push it back just as swiftly.

This was not just a freak of Darwin's finches. Naturalists are now documenting similar reversals of fortune elsewhere in nature as well, including populations of Darwin's "imps of darkness," the marine iguanas of the Galápagos. The iguanas forage for seaweed in the shallows and then bake in the sun to digest it, while finches hop over them and sometimes pinch a fly or two from the lizards' foreheads and dragon crests. It is hard to imagine neighboring animals with lives that are less alike, and yet for the evolution of the lizards the pressures of the droughts and the flood appear to have been countervailing too.

Most of us think of the pressures of life in the wild as being almost static. Robins sing in an oak tree year after year. We imagine that life puts more or less the same pressures year after year on the robin and the oak. But the lives of Darwin's finches suggest that this conception of nature is false. Selection pressures may oscillate violently within the lifetimes of most of the animals and plants around us, so that the robin must cling to the oak, and the oak to the ground, in chafing and contrary winds. It is as if each living thing on earth is holding on at the very shore of an ocean, in rough and invisible seas, swaying in place as each wave shoves it toward the shore and then tottering as the broken surf drags it back again.

The stutter-step quality of the action is yet another reason that natural selection has been missed in most studies of live populations in the wild. If you measure natural selection over the course of a whole generation you may miss the many slings and arrows that it has taken along the way, the conflicting pressures in the nest, in the first days out of the nest, and on the yearlings and the adults; or on the acorn, the green shoot, and the towering oak. Each stage of life may have experienced an intense episode of natural selection, and yet their effects may have obscured each other's traces by the time the very last of the generation has shuffled off the earth. Species of animals and plants look constant to us, but in reality each generation is a sort of palimpsest, a canvas that is painted over and over by the hand of natural selection, each time a little differently.

When the finch watcher Jamie Smith left the Galápagos and began watching the sparrows on the island of Mandarte, in British Columbia, he did not know if he would find selection events there. His study provides a good basis for comparison because the sparrows and finches are closely related, and in some ways the situation of the song sparrows on Mandarte is much like that of Darwin's finches on Daphne. There is a small resident population of sparrows. They are there year round. They do not migrate.

Much of the work that Smith and others have carried out since the early 1970s on Mandarte is similar to the work that *El Grupo Grant* has done on Darwin's finches: catching and banding sparrows, measuring their beaks and wings, following their fates.

A few years ago, Smith was working on a paper on natural selection in these song sparrows, looking at the same traits as in Darwin's finches. His major result was there was no evolution in the birds. Smith was going to report no selection in song sparrows.

Before Smith published this report, another veteran of the Galápagos finch watch, Dolph Schluter, joined him at the University of British Columbia, in Vancouver. Smith told Schluter that natural selection was not doing anything to his sparrows.

"So, I didn't believe it," Schluter remembers now, with a laugh. Schluter was fresh from the Galápagos at the time, and he was full of the power of natural selection. "Jamie told me, 'Fine, here are the data. See for yourself.' "

Dolph took a look. He knew that Smith had checked for evolutionary trends by comparing a generation of sparrows at birth and the same generation at death. Dolph decided to look at the birds year by year instead. He also broke each year into three components, studying the young sparrows' survival rates in their first year of life, as they weathered their first Canadian winter; their survival as adult birds in each succeeding winter; and their success in rearing offspring in each breeding season.

When Schluter put the sparrows under the microscope in this way he found that natural selection had been working quite ruthlessly among the sparrows.

Among the males, selection had worked to eliminate the outliers—the birds that deviated most toward large or small. This is what is known as stabilizing selection. This kind of selection pressure helps to explain why the sparrows on the island are so much less variable than the finches on Daphne.

Among the females, Schluter found oscillating selection, and the case was remarkably similar to the one on Daphne Major. There had been two tremendous population crashes in the course of the study, just as on Daphne. One crash was caused by bitter cold weather, high winds, and snowfall, during the winter of 1987–88. The watchers on Mandarte took a census before and after the cold spell. (It would have been tough to get out to the island in the middle of a cold spell: the ride out in a Zodiac takes about one hour, and in bad weather you arrive at the island covered with ice and snow.) When the sparrow watchers arrived after the cold spell was over, they found that they were down to eight birds (which was scary for Smith; his flock almost went extinct). The second population crash was not caused by a hard winter—in fact, the sparrow watchers still do not know what was killing their birds. But the females were pushed one way by the first crash, and the other by the second, again much as on Daphne Major.

"My result: lots of selection," says Dolph merrily. "At least one event every year." Yet when he summed all these changes over the lifetime of a generation of sparrows, he saw no selection at all, just as Smith had said.

"So we were both right," Dolph concludes. Summed over years, the effects of natural selection were invisible. But at each stage of their lives and each year of their lives the sparrows on that little island had been "daily and hourly scrutinized" by the hand of natural selection, much as Darwin imagined, only in fast motion.

The population on Mandarte is still being pushed every year, first left, then right. Smith and his team have not progressed as far as the Grants in determining the causes behind those pushes. They have never tried to count the seeds and bugs on the island and match them with the numbers of the sparrows, for instance. (Mandarte is much more complicated than Daphne.) But year after year they are seeing fluctuating selection at different life stages—opposing selection, between young and old stages of life, just as in the finches. And they are seeing oscillating selection from one year to the next, also as in Darwin's finches.

"You start to view species not as constant entities but as fluctuating things," Dolph says. "A species looks steady when you look at it over years—but when you actually get out the magnifying glass you see that it's wobbling constantly. So I guess that's evolution in action. The world is not as stable as you think!"

What seemed most striking to Dolph as he studied the selection events hidden in Smith's data was the boring uniformity of Smith's birds. Compared with Darwin's finches, the sparrows of Mandarte might have been turned out by a cookie cutter. They showed only the very slightest of variations from one bird to the next in length of beak and length of tarsus. Yet even these variations, trivial as they seemed, had helped to decide who lived and who died. "That is pretty amazing to me!" Dolph says. It means populations don't need to be excessively variable in order to experience natural selection.

"Selection doesn't happen just in the Galápagos," Dolph concludes. "It happens in your backyard."

WHEN DARWIN LOOKED at the fossil record, he saw it static and frozen for long stretches. In retrospect that is not surprising. If selection events do not show up among a tribe of sparrows when summed over the course of a single generation, consider how much less visible these events will be in the strata of rock beneath our feet, in which the generations have been summed for many millions of generations.

One of the most famous examples of "rapid" evolution in the fossil record is the ancestral lineage of the modern horse. That was the subject that Huxley chose for his lecture "The Demonstrative Evidence of Evolution." The transmutation of an upper molar in *Mesohippus* from *Hyracotherium* is one of the most rapid changes whose sequences are preserved like a snip of flicker-frame silent movie in the rock record. This event took about one and a half million years, or about half a million generations.

Even that rate is molasses and stasis compared with the stutter step of live finches and sparrows. Compared with life above the ground, the record of life in the stones flows almost as slowly as stone itself. The apparent lack of action in the fossil record confirmed for Darwin his belief that evolution by natural selection must be a rare event, and that any and all action must be unendurably, geologically slow. It is because the changes in the record are invisible, and the changes around us were at the time invisible, that he says in the *Origin* we "see nothing of these slow changes in progress, until the hand of time has marked the lapse of ages."

In 1949, the evolutionist J. B. S. Haldane suggested that the rate of evolution be described in terms of a universal unit, whether the change took place in the animal or vegetable kingdoms, among the living or the long extinct. This unit he named in the same spirit of tribute that we call continents and oceans after their discoverers. He called his unit the darwin.

To keep things simple, Haldane said, let us define the darwin as a change in the length of some character. Since we want a universal unit, and not one that depends on the size of the creature, let us use the percentage change. The beak of a duck-billed dinosaur might lengthen by a few meters in a million years, and a bird's beak by a few millimeters in a million years, Haldane said, and yet both could represent a change of 10 percent. A percentage change will mean the same thing in myriad living and once-living forms, from the beaks of birds to the teeth of hyenas, from the skulls and cannon bones of horses to the inner whorls of ammonites. Haldane defined one darwin as a change of 1 percent per million years.

When Haldane looked at a few representative fossil records of evolution he found that the rate of change was very slow, on the order of 1 percent per million years, or a rate of change of a single darwin. Other investigators since Haldane have borne out his guess that this snail's pace is typical of the fossil record.

Haldane concluded as Darwin had that the rate of evolution by natural selection in the world around us must be infinitesimally slow, far too slow to watch, that it could only be watched in the long, slow additions of the fossil record. Rates in the living world would have to be measured in millidarwins. In artificial selection, he said, you could get rates of thousands of darwins, but that is not something you would see in the wild: "Rates of one darwin would be exceptional in nature."

"Such calculations are extremely rough," Haldane said, "but they suggest the remarkably small order of magnitude of the selective 'forces' which are at work if natural selection is largely responsible for evolution, and the extreme difficulty of demonstrating them in action."

Now it is possible to translate the evolution of Darwin's finches in the drought and the flood into Haldane's whimsically named unit, the darwin. In the drought the change was 25,000 darwins. After the flood the change was about 6,000 darwins.

So there is an enormous gulf between what we see when we take the time to watch the living world in action, and what we see when we look at the world recorded in stone. To set this discrepancy in a broad perspective, an evolutionist not long ago compiled more than five hundred cases of evolutionary change, from short and fast experiments in artificial selection (events that took months or at most a year and a half) to evolutionary experiments in the fossil record (events that took millions of years). The evolutionist, Philip Gingerich, translated all of these events into darwins. He discovered a simple pattern, a pattern that is just the opposite of what earlier evolutionists—the whole lineage of evolutionists from Darwin to Haldane—would have expected. The closer you look at life, the more rapid and intense the rate of evolutionary change. The farther back in time you stand, the less you see.

In a single year, you can find rates of change as high as 60,000 darwins. But in the fossil record the average is only a tenth of a darwin.

The reasons for the discrepancy are not far to seek. If at any time in those millions of years a species changed swiftly, but the rest of the time it changed slowly, that start-and-stop motion would average out into a very sluggish movement. What is more, if the species changed first one way and then the other way, over and over again, as Darwin's finches did in the first decade of the Grants' watch, then the fossil record would register virtually no change, a near equilibrium. Yet the beak of the finch is in fact in so much evolutionary motion that as soon as people started watching closely enough, they saw it change right before their eyes.

"There's quite a bit of wobble in the fossil record too," says Peter Grant. "But you don't normally see it. When you look at a series of fossils, the usual practice is to take 3,000 or 5,000 or 10,000 years as the unit. You take the average position at the beginning and the end, and average the rate of change in between.

"Even then, you see wobbles in the record. It's one in a thousand wobbles, if you like. You rarely get successive generations preserved in fossils. It's like looking at specimens of Darwin's finches from 1874, 1932, and 1987. You may get some wobbles there—but you miss all sorts of action in between."

The fossil record is just too primitive a motion-picture camera to capture the fast-moving life. Rapid motion disappears like the whir of a hummingbird's wings. In such a record, the two wonder years of Darwin's finches would disappear as surely as a wing-beat up and a wing-beat down, canceling out in the blur.

If we look at the billowing smoke of a volcano from close up, we see intense and rapid motion, enormous and dangerous turbulence. If we look at the eruption from far, far away (a safe distance that puts it almost to the horizon), the smoke seems to hang in the air almost motionless: we have to watch a long while to see any change at all. The evolution of life turns out to be rather like the eruption of a volcano. The closer you look, the more turbulent and dangerous the action; the farther your remove, the more the living world seems fixed and stable, hardly moving at all.

It is amazing to think of all these species around us not fixed but in jittery motion. It is like the difference between the old view of solid physical matter around us—the view in the time of Newton—and the view now, of infinite motion down to the level of each atom and molecule, and below, in the ceaseless assault and battery of elementary particles. The beak of the finch is an icon of evolution the way the Bohr atom is an icon of modern physics, and the study of either one shows us more primal energy and eternal change than our minds are built to take in. Yet like the vista of the atoms, the vista of evolution in action, of evolution in the flesh, has enormous implications for our sense of reality, of what life is, and also for our sense of power, of what we can do with life.

"This jittering is an aspect of all populations everywhere," says Dolph Schluter. "It is demonstrating that populations are currently dynamic—still wobbling—so that a larger change in the environment at any moment might push them one way or the other." If they weren't jittering, that would suggest that the processes that brought them here had finished, that the creation was over, just as the universe would be moribund or dead if there were no motions to be found in its atoms. But these motions are to be found everywhere, Dolph says. "They are ever-present."

Having seen even this much of the view from the rim of Daphne Major, we can no longer picture the story of life as slow and almost static, a world view for which the chief emblem of evolutionary change is a fossil in a stone. What we must picture instead is an emblem of life in motion. For all species, including our own, the true figure of life is a perching bird, a passerine, alert and nervous in every part, ready to dart off in an instant. Life is always poised for flight. From a distance it looks still, silhouetted against the bright sky or the dark ground; but up close it is flitting this way and that, as if displaying to the world at every moment its perpetual readiness to take off in any of a thousand directions.

PART TWO

New Beings on This Earth

We will now discuss in a little more detail the
struggle for existence.

— CHARLES DARWIN,
On the Origin of Species

Chapter 8

Princeton

These birds are the most singular of any in the archipelago.

— CHARLES DARWIN,
Journal of Researches

... the more you look the more you see.

— PETER GRANT,
*Ecology and Evolution of
Darwin's Finches*

Mid-morning and mid-June in Princeton, New Jersey. Rosemary Grant is at work in her office, Eno Hall, Room 106. She wears an Icelandic sweater, a long blue Laura Ashley dress, and sandals. Sunlight slants in the bay window behind her, through the spreading green arms of a katsura tree with shaggy bark.

She is sitting on a Danish backless chair, at a table with wrought-iron legs. This table runs the length of the room and holds a Casper GM-1230 computer, which is an IBM clone; a Hewlett Packard LaserJet printer; a second printer; and a Macintosh II with a screen that would be large enough for a living-room television.

The Grants got back from the islands only a few weeks ago. Already they have transcribed every one of the numbers from their waterproof notebooks into the computers. Now they plan to spend a year without teaching, alone with all their years of records. "We won't be working," Peter tells people at Princeton. "Our purpose in taking this sabbatical is not to work with our hands but to creep away and analyze our data."

"Why do you say we won't be working?" Rosemary protests. "That's work."

"Yes. In some ways that's the real work, and everything else is play."

Rosemary runs her finger through the air above the tidy row of boxes that she and Peter keep shelved above the computer table. Charles Darwin kept his notes in thirty or forty big portfolios, which he stored in his study on specially built racks to the right of the fireplace. Thanks to his powers of organization and his capacious mind he managed to file and retrieve an extraordinary breadth and depth of information in those portfolios. But he could not have handled as much data as Peter and Rosemary keep in the rows of little boxes on this shelf, the archives of the International Finch Investigation Unit.

Rosemary chooses a box, slips out a diskette, and pops it into the Macintosh. "Okay," she says, after a brief search. "Here's 3425." Number 3425 is one of the two rogues that Rosemary caught half a year ago on Daphne's north rim. He was the first of the two cactus finches that hopped into her traps. The rows of numbers on the screen summarize what the Grants and their assistants have learned of the fortunes of 3425 since that morning in January.

"So far this year . . . he had two clutches," Rosemary says, speaking slowly as she decodes the numbers and letters on the screen. "And he had the same female for both clutches, 5582. In the first clutch, there were three eggs. Two of them hatched, and only one fledged. In the second clutch, three eggs hatched . . . but again, only one fledged."

The drought is over, which is why 3425 has been so busy. The first bout of rain fell on Daphne in February. Within days the *Portulaca* flowered, and the torchwood trees put out leaves and filled the air with the memorable odor of their greenish-white blossoms. Tall grass sprouted and covered all the paths. Even after all these years, the speed of Daphne's metamorphosis still took the Grants by surprise.

The finches bred and bred. Peter and Rosemary and this year's assistants ran up and down the sides of the volcano, trying to keep up with hundreds of nestlings—Darwin's latest finches. Then the Grants climbed down to the welcome mat and into the *panga* of a little local boat called the *Flamingo* to visit the finches of Gardner, Floreana, Genovesa, and Española.

Just off the island of Floreana the *Flamingo*'s engine failed. An engine cable had broken, and the captain went belowdecks to repair it. The *Flamingo* drifted toward the resounding surf of a rock called Enderby. The swell was long and deep, and it pulled them closer and closer to the boiling surf and the rock. Rosemary, Peter, and Thalia stood at the railing waiting for the engine to start, watching the rock get closer and closer. At the last second the captain got the cable fixed and the engine started; another season of precious numbers made it to this shelf in Princeton.

Rosemary ejects the diskette from the Macintosh and loads another one, labeled "NESTOTAL76-91." There is a long pause while the computer clicks and clacks, but the screen stays blank. "This is a big file," Rosemary says while she waits: "5,575 kilobytes, I think it is." A file that size could hold about a million words, or the complete manuscript of Darwin's "Big Book," *Natural Selection*, plus several editions of the *Origin* and the *Descent of Man*. That is how much information the Grants and their field assistants have compiled on the nestlings born to Darwin's finches on Daphne Major from 1976 to 1991.

Abruptly, columns of numbers cascade down the screen and fill it from top to bottom. "That's why it took so long," Rosemary says, nodding at the dense little rows of numbers. "And this is just the *start* of the file." She scrolls down through the streams and streams of numbers. "Okay, here he is again. Our 3425 is quite an old male. These are all the times that 3425 has bred in his life: one, two, three . . . ten times. For the first time in 1982. Then he bred"—she counts aloud—"*eight* times in 1983." That was the wonder year, the year of the flood. "He bred again in 1984, once. That's about as much as *anybody* did in '84. . . . Then he did not breed again until this year."

With a few strokes of the keyboard, Rosemary can also locate all of the females that this old finch has mated with: the life and loves of 3425. Rosemary puts her index finger to her cheek, studying the screen. "He bred in 1983 with the same female as in '82 . . . seven times. But in the middle of the season he had another female, 4629." Like so many other finches in the flood year he turned fickle and swapped mates. "She died in 1987. And in '84 he bred with yet another female, 5538. She's also dead now."

Peter pokes his head in Rosemary's door. In place of the black binoculars he wears in the Galápagos, a pair of black reading glasses hangs by a cord around his neck. He is deeply tan, and dapper in a tropical khaki shirt. If anything, he looks even more like Charles Darwin here at Princeton than he does in the islands. Some of his colleagues wonder if he nurtures the resemblance. As if to deflect this suspicion, Peter has taped to his office door not one but two identical yellowing newspaper photos of a bearded Scottish bagpipe player. Peter looks even more like the bagpipe player than he looks like Darwin.

"How's it going?" Peter stoops and glances at the monitor, which is still displaying the love life of 3425, with the running heads above the columns, SPECIES, EGGS, NESTLINGS, FLEDGLINGS, SECTOR. . . . He straightens briskly, without studying the screen, and pronounces with a positive pleasure: "I make it ten fledglings produced over his lifetime before this year. And this year he produced two. And none of those were recruits."

"Oh, you have that in your head, do you?" Rosemary says.

"I'm practically certain of it," says Peter. "Swapping mates didn't do him any good. Of all the fledglings he's produced—two of the ten in 1984, and the other eight in 1983—not one has reproduced. So this bird is a *loser.*"

Rosemary casts up onto the screen the lifetime batting average of another old male of the same generation, 2666.

"He's been a good producer of fledglings, 2666," Peter says, this time without even looking at the screen. "In the region of thirty. And some of them are still around and breeding. At least two. Maybe better than that."

"He wasn't the *most* successful breeder, though," says Rosemary. "That was 720—was it he, Peter? With ten recruits. Quite an achievement."

"Ah, yes, 720. God bless his soul; rest in peace! The most successful of all time."

The sun has gone behind a cloud now. A high of 87° F is predicted for today, but in Rosemary's office the air is cool. Tote umbrellas and raincoats hang from a peg rack by the door. Otherwise there is very little personal in the room: only the row of computers, the shelves of diskettes, a great wall of books, a few framed portraits of cactus finches, and Nicola Grant's phone number taped to the telephone.

Now that the girls are grown, Rosemary's role in the research has expanded. The first monograph of the finch watch, published in 1986, is by Peter R. Grant. The second monograph, published in 1989 and focusing on Genovesa ("Rosemary's Island"), is by B. Rosemary Grant and Peter R. Grant. "They work as a couple—they work as a unit," says one old friend. "The world will take much of what they do as Peter. Yet they really do something that transcends either one of them.

That's developed over a long period of time, probably to a greater point than even they themselves are aware."

Peter vanishes, then pokes his head back in Rosemary's door: "I'm setting up lunch, darling." But Rosemary keeps going. She calls up the life story of 5608, which was the other rogue finch on the north rim that morning: the Princeton bird, banded orange over black. Another loser; but then, the losers are so much more common than the winners.

"So," she says at last. "For any bird, if we want to know the dimensions of its beak, or who were its parents, or when was it born, or where was it born, or who were its siblings in the nest"—she recites this list almost in a tone of incantation—"we have them all right here, four or five generations, going right back." That is, four or five generations counting by life-spans, more than twenty generations counting by begats: a Book of the Chronicles, or a Book of the Kings, all devoted to Darwin's finches.

"Then we have the *vegetation*," Rosemary says, with a little laugh, waving her hand at another set of little boxes. "It's all there. Here are the cactus data, for example." There is the *Star Trek* sound of a diskette ejecting from the Macintosh, and then the thunk of "Compiled Cactus Data" going in. By pressing a few more keys she can produce an instant graph of cactus yields on Daphne Major for the last fifteen years.

"And then the *seeds*, the yearly seeds," she says. "We have seed files across all the years up to this." Rosemary loads and displays one of these disks, an enormous file called SEEDYEARN.

"Then of course there's *rainfall, temperature. . . .*" More boxes of diskettes. "And we have the notebooks, of course, and the daybooks." (The daybooks are fair copies of the notebooks, made in the evenings on the islands.) She riffles one of the latest daybooks: pages and pages of neatly inked numbers.

"And these are the *songs*." This morning the songs of all these generations of Darwin's finches are neatly stacked on a table beneath Rosemary's bay window. Students have transcribed the songs from tape recordings using a machine called a sound spectrograph. The machine takes each song on the tapes and represents it in rows of thin, finely spaced vertical lines. Each picture is a snapshot-sized piece of white paper, bearing a single finch's song.

"We're getting them bound, just today," says Rosemary, flipping through one stack of songs. She laughs again. "And so on, and so on, and so on! It all takes a surprising amount of time, but it's fun. As of this moment, every finch on the island has been banded but one. That one is right up at the top of Sector Four. He's been there for years. A terrible place, really horrible. Everyone has tried to catch him. He's right up here," she says, pointing to the spot on a map of the island. "It's very steep, going down into the crater. Impossible to put a net there. Also he *flies*. He's a *very* difficult bird."

THE GRANTS HAVE CARRIED the islands home in their pockets. Now, sitting in Princeton, they can look for action that was invisible from the rim of Daphne Major: secrets they were too close to see, when they were measuring the beak of one finch while three others perched up and down their wrists to watch.

Although their sabbatical has just begun, and they are still taking turns at the computers, checking and rechecking their data, they feel an excitement they could not explain very clearly to anyone else. Some of the latest figures from the field are puzzling. Peter and Rosemary have found a set of statistics that does not fit, the suspicion of a surprise. They first noticed this anomaly while they were out in the islands. In fact they have been half noticing it for a few years. Now the more they study it, the richer and stranger it seems.

Back in 1983, for instance, during the mating rush of the super Niño, a male cactus finch on Daphne Major, a *scandens*, courted a female *fortis*. This was a pair of truly star-crossed lovers. They were not just from opposite sides of the tracks, like the Prince and the Showgirl, or from two warring families, like Romeo and Juliet: they belonged to two different species. Yet during the chaos of the great flood they mated, and they produced four chicks in one brood.

"For most families of four," says Peter, "you get zero, or one, and the rest are never seen again. Maybe occasionally two. In this case all four fledged: 5626, 5627, 5628, and 5629. The first three were sisters, and the last a brother. The brother bred in one year and then disappeared. But the sisters have done extremely well."

"And their offspring have bred," says Rosemary.

"And their offspring have bred," Peter agrees. "I think 5629 turned up in an owl pellet. But they all bred."

"Here they are," says Rosemary, opening a file called HYBRIDNEST. "Born in 1983. Their father was a *scandens*, 4053. Their mother was a *fortis*, 1536. Three sisters and a brother."

Rosemary opens the brother's file first. "Okay," she says. "The first time he bred, he was the father of four eggs, which produced three fledglings." So the brother was off to a strong start before he vanished into the owl.

Next Rosemary opens the sisters' file, and the numbers rush down the computer screen like an invading army. "Now look at this," she says. "We have some naughts in a few of these years, but if you add up the totals . . . forty-three grandchildren from the three sisters! As of this year. That does not make them the all-time champions on Daphne Major, but it puts them high. The four were born only in 1983, and they have forty-six grandchildren altogether. That's very high.

"If I follow those through, there will be great-grandchildren. There'll be a lot of great-grandchildren, because I know a lot of those forty-six went on to breed. Some of them bred even in 1983! So we're going to add up to a *lot* of birds.

"Goodness! This year, 5626 had six eggs. She produced six eggs and six fledglings. And 5627 produced six eggs and five fledglings. And 5628 produced five eggs and four fledglings."

According to conventional wisdom, this should not be happening. When David Lack was in the islands half a century ago he tried very hard to spot a finch of one species pairing off with a finch of another, and he did not detect a single case. None of the ornithologists who visited the islands before Lack had ever reported such a thing either. Lack shipped a multitude of caged finches to San Francisco, where Robert Orr, of the California Academy of Sciences, tried to breed them. The birds mated in San Francisco, but only like with like, only within their own species. "Clearly hybridization between species is rare, if not absent," Lack concludes in *Darwin's Finches*.

It was the Grants' Finch Unit that spotted the first hybrids among the finches. Peter Boag and Laurene Ratcliffe made the discovery in 1976, the year before the great drought. By then the finch watchers knew enough birds that little oddities and rarities could begin to stand out. In that one year Boag and Ratcliffe noticed five male *fortis*, medium beaks, mating with five female *fuliginosa*, small beaks. Altogether these five couples produced a dozen eggs. None of the fledglings made it, of course—not a single one of the birds in 1976 survived the drought of 1977. But when the survivors of the drought of 1977 paired off, one of the new couples on the island was, once again, a male *fortis* and a female *fuliginosa*.

In nature, the isolation between sibling species is seldom absolute. Among closely related plants, hybridization is more the rule than the exception, as Darwin notes: even annuals and perennials, or deciduous and evergreen trees, "can often be crossed with ease." Among animals crossing is less common, but it happens. Mallard ducks and pintail ducks interbreed freely. In zoos, lions and tigers can be made to hybridize and produce tiglons. Zebras and horses can hybridize and produce striped sterile zebroids.

If two species intercross very rarely, the mixing will not dramatically alter their gene pools. Often their offspring will be less fit than purebred offspring. Horses and donkeys interbreed all the time, and some of their offspring are famously sturdy, hardier than their parents. But the hybrids are unfit in the technical sense, the sense used by evolutionists, who define and measure fitness in terms of numbers of offspring bestowed on future generations. Crossing a female horse and a male donkey will never change the gene pools of horses and donkeys by even a single gene, because all mules are sterile. Crossing a male horse and a female donkey will not change their lines much either, because hinnies are small, and less useful around the farm; so farmers seldom let them breed, though not all hinnies are sterile.

That was the aspect of hybridization that interested Darwin. His chapter on "Hybridism" in the *Origin* is mostly about hybrid sterility. He assumes most species would be sterile if they interbred, "for species within the same country could hardly have kept distinct had they been capable of crossing freely."

The Grants have always assumed, with Darwin, that the offspring of their Galápagos hybrids must be relatively unfit. They thought the hybrids would be rejected, culled, weeded out by the hand of selection. So would the tendency to interbreed. The rare and peculiar tastes that lead to a star-crossed marriage would remain peculiar and rare. For if the products of these mixed marriages had any advantage in the struggle for existence, then the tendency to interbreed would be favored, and hybridization would become more common. And if it became common, and stayed common long enough, all of these species would fuse. The family tree of thirteen finches would taper to a single twig. Their celebrated beaks, their ingeniously varied tool kit, would become one beak.

Since there is not just one species of Darwin's finches in the Galápagos, Rosemary and Peter were positive that the hybrids and their experimental beaks must be at a disadvantage. And they thought they knew why. Everything they had learned about the struggle for existence of Darwin's finches in Darwin's islands argued that the hybrids should do badly. They had seen again and again that the slightest variations in the beaks of these birds can change their fates. Half a millimeter can decide who lives and who dies. Since these slight variations are passed down from one generation to the next, the brood of a small beak and a medium beak would be likely to have intermediate beaks, equipment that would sometimes differ from their parents' not by one or two tenths of a millimeter but by whole millimeters, maybe by many millimeters. If two hybrids mated, their nest would be a mother's nightmare, a little novelty shop of strange, assorted beaks. Daphne Major is not a forgiving place. A line of misfits should not last.

During the first half of their study, the years before the great flood, hybrids were so rare that it was impossible for the Grants and their team to draw any definite conclusions. But the mixed bloods did seem to be doing less well than the purebred birds.

That is why the Grants are so puzzled now. There have been too many cases to put down to chance. On Daphne, one of the Grants' favorite dynasties began some years ago when yet another *fuliginosa* female paired off with a *fortis*. "She was the smallest bird on the island," Peter says. "She's the very smallest *fuliginosa* we've ever had on the island. And she has outproduced all others of her kind, breeding with a *fortis*. Never paired up with another *fuliginosa*."

This tiny *fuliginosa* was Number 006. She paired up with *fortis* Number 459. They nested along the inside wall of the crater, near the top, just where it begins to slope down, in the territory the Grants call "Downstairs." "And they just, you know, bred very well," Peter says. "Some of the birds we have on the island right now are the grand- or even the great-grandchildren of that pair."

The misfits are not dying out; they seem to be thriving. In place of a few interesting individuals, Peter and Rosemary are seeing outstanding families. During their ten-year watch on the island of Genovesa ("Rosemary's Island") some of these hybrid families were like Genovesa royalty. They turned out fledgling after fledgling, recruit after recruit, when families all around them were dying out and disappearing. "It went on for three generations!" says Rosemary.

IN ENO HALL there are more distractions than on Daphne Major. Even on sabbatical the work is stop and go. Peter and Rosemary often work through lunch, sitting at a desk in the corner of Peter's office, jumping up now and then to scrawl on the blackboard, then back to their posts with the door ajar between them. It makes an island of a sort, but there is still the telephone.

"From New Zealand," Peter calls through the door. "They didn't receive the fax."

"From North Carolina Biological. I told them you placed the order."

By fits and starts they analyze their numbers, compiling and debugging the data sets, collecting all their records on hybrids since the beginning of the great flood. Peter has to fly to Hungary to give a paper at the International Congress of the European Evolution Society in September, and he and Rosemary scramble to finish the last of the calculations before he goes.

They compare the hybrids' survival rates to the rates of the normal birds in their cohorts. They compare the hybrids' success in the nest. They compare the successes and failures of several generations, one against the next.

"Hey, we've got it!" Peter cries at last. The grand totals are as weird as they thought.

Before 1983, the year of the flood, pairs of medium beaks and small beaks on Daphne Major produced thirty-two fledglings. None of these hybrid fledglings bred at all until the flood. In fact, only two of them lived long enough to see the flood. These misfits were unfit.

But *since* 1983, which can truly be called a watershed year, the hybrids have done better. Those that hatched after that year were more likely to survive than the antediluvian hybrids, and they were more likely to breed. They were also slightly more successful than the offspring of purebred *fortis* or *fuliginosa* pairs. And these odd couples went right on producing for the rest of the 1980s.

In all demographic studies, one of the crucial statistics is the rate of replacement: the accounting of births versus deaths. There are always three possibilities. A cohort can replace itself over time, births keeping up with deaths; the cohort can grow, births exceeding deaths; or the cohort can dwindle, deaths exceeding births.

When the Grants look at the very latest of this year's breeding statistics for Daphne, they can see that the purebred *fortis* and *fuliginosa* born in 1987 have not managed to replace themselves. They have not produced enough fledglings to keep up with the rate at which they themselves are disappearing. As the Grants put it, the starting number of fledglings in 1991 is smaller than the starting number of fledglings in 1987.

But the *crosses* between these two species are doing much better. The hybrids born in 1987 have more than replaced themselves: they have increased by a factor of 1.3.

Crosses between *fortis* and *scandens*, the medium beak and the cactus finch, are doing better yet. For these hybrids the contrast with the years before 1983, the antediluvian years, is striking. Before the flood, there was only one such pair on Daphne. This pair produced only one fledgling, and their fledgling survived less than one year. But by the end of the breeding season in 1987, the Finch Unit had discovered and monitored a total of five different pairs of *fortis-scandens* hybrids, and the five pairs had produced twenty-three fledglings. Many of those fledglings lasted longer than purebred offspring of either species; they even outlasted the *fortis-fuliginosa* hybrids.

Something has changed since the flood. Something is happening. Strange as it seems, these hybrids are the fittest finches on the island. If the Grants' hunch is right, this is a missing piece of a puzzle, a puzzle they have been struggling with for the last twenty years: "that mystery of mysteries," the origin of species.

Chapter 9

Creation by Variation

The Galápagos seems a perennial source of new things.

— CHARLES DARWIN,
letter to Joseph Hooker

"Evolution happens the whole time," Peter Grant tells his classes at Princeton. "You look up in surprise. But evolution is always happening. Completely contrary to Darwin's view that very, very slowly, very intermittently, life evolves.

"Geneticists will tell you that. Evolution is always happening. What they mean is that the genes of this generation are not precisely what they were the preceding generation. Nor will they be precisely the same in the next. And evolution is that change. And it is almost a certainty, a mathematical certainty, that the genes will never be the same.

"Darwin was right in a sense: you won't see it. You look out at the maples on this campus, or the robins, or the gray squirrels, and year after year they look the same. They aren't. They're different. But you can't see it, the differences are too subtle."

Peter sometimes thinks about the time that Darwin himself almost saw it happen. Darwin was a man of excruciatingly regular habits. Year after year, twice a day if health allowed, Darwin used to leave his house, cross to the farthest end of the kitchen garden, and go out through a wooden door in the back hedge. Then he would stride down a fenced path between two lonely and—in the wintertime—desolate meadows to a strip of land that he rented from his neighbor, a wealthy astronomer.

Back there, with the astronomer's permission, Darwin had planted trees and laid out a trail among them. This was his Sandwalk, his Thinking Path. The ground was littered with flints, which looked to him when he first saw the place "like great long bones." At the start of his walk, Darwin would collect a heap of these flints and set them on one side of the path. Then as he went around and around he kicked away a flint or knocked it aside with his iron-tipped stick, to count the laps.

Year after year, day after day, the Sandwalk was evolving around him, though Darwin did not see it. One cold and dreary March, for instance, when the whole household was sick with "coughs & colds & rheumatism" and "Hooping cough," Darwin noticed that the birds in his trees had been decimated by the cold. In the third chapter of the *Origin*, "Struggle for Existence," he estimates that the freezing weather had killed about four-fifths of the birds on his grounds: "and this is a tremendous destruction," he writes, "when we remember that ten per cent. is an extraordinarily severe mortality from epidemics with man."

This is the event that intrigues Peter. By then—it was the winter of 1855—Darwin was already drafting his "Big Book," *Natural Selection*. Yet Darwin did not dream he could see his process in action among the birds in his own backyard, so he did not try.

"I really do believe that if he had been thinking along the lines that selection can be observed, he would have done something about it that winter," Peter says. "I believe he came very close to documenting the process."

If the weather was bad enough to have knocked birds out of the trees, which sometimes happens in England, and New England, then all Darwin had to do was get his gardener to collect some of the dead blackbirds.

"If he *had done*," Peter says, "Darwin might have noticed that some were big, or fat, or long in the beak, or long in the leg, or weird in some way." Then he could have asked his long-suffering butler Parslow to shoot some of the birds that had survived the winter, so that Darwin could compare and contrast the measurements of the dead and the living.

"He might not have found anything," Peter says. "But it's coincidental that about four-fifths of the birds died. That's how many disappeared the first time we were watching—85 percent in the 1976–77 long drought.

"Possibly those blackbirds could have done it for him."

This is what evolutionists have been doing ever since Darwin. They have been walking around and around his Thinking Path, kicking the flints, looking for ways to test and extend the ideas that Darwin worked out on those quarter-mile laps. Darwin's followers are always thrilled when they can confirm what Darwin saw, and they are even more thrilled when they can see what Darwin missed. Probably no other major branch of science today is so haunted, dominated, and driven by the thoughts of one man.

The Grants are retracing the path of Darwin's thought by beginning where Darwin began—in more ways than one. Not only do they gather their data in the center of Darwin's islands. They gather it in the same simple, concrete, objective units that Darwin considers the origin of all evolution: individual variations, measured in millimeters. Every evolutionist appreciates the value that Darwin attached to variations, but not many people in the field are as excited by variation as Peter Grant. When he lectures, Peter sometimes sounds even more interested in variation than in evolution itself. "Evolution *is* change in variation," Peter says. "By studying evolution in action we can better understand the variation." He draws a bell curve on the blackboard and then gestures, as if to lift the curve off the board into the air with his hands, trying to suggest to his students the dimension of time— coaxing the curve off the wall with his hands and warping it eloquently as it floats out into the air. "*This* is what we have to explain," he says, pointing at the curve on the blackboard. "This is variation. And its change through time."—pointing with his eyes at the path of the curve through the air—"is evolution."

WHAT MAKES new species? How, exactly, does variation lead to creation? Why does a line of life keep more or less the same shape and more or less the same habits for thousands, tens of thousands, sometimes millions, of years, and then branch out into new lines, new shapes, new habits? This is what Darwin wanted to know, what the Grants want to know, what Darwin's followers have been debating ever since Darwin.

It is one thing to demonstrate, as the Grants have done, that natural selection leads to evolution. It is another and much more complicated thing to demonstrate precisely how this evolution leads to new species; and despite the title of his greatest book, Darwin himself never spells out the details.

Darwin calls the *Origin* "one long argument," and this is the step in the argument that many of his readers find the hardest to follow, the step that feels like a leap of faith: from slight individual differences in one nest, or one seedbed, or one family album, to the striking differences between species. They can accept that Darwin's mechanism, natural selection, can refine adaptations. They can understand that Darwin's process might play, in this way, a sort of supporting role on the stage of life, as an editor of beaks and bodies, an improver of lines. But they cannot believe the process can create something new.

Not even Darwin's friends were satisfied that natural selection leads to the origin of species. Hooker, the botanist, wrote to him diplomatically:

> You certainly make a hobby of Natural Selection, and probably ride it too hard; that is a necessity of your case. If the improvement of the creation-by-variation doctrine is conceivable, it will be by unburthening your theory of Natural Selection, which at first sight seems overstrained—i.e., to account for too much.

Huxley, Darwin's standard-bearer, managed to give a sort of victory speech for evolution, "The Coming of Age of the Origin of Species," without mentioning natural selection once.

After Darwin's death, many biologists found it easy to accept evolution and impossible to accept Darwin's chief explanation for it. Evolution, yes; selection, no. William Bateson, the founder of modern genetics, wrote an elegy for Darwinism in 1913, calling it "so inapplicable to the facts that we can only marvel . . . at the want of penetration displayed by the advocates of such a proposition."

Nordenskiöld's *History of Biology* dismissed Darwinism forever in 1924:

> To raise the theory of selection, as has often been done, to the rank of a "natural law" comparable in value with the law of gravity established by Newton is, of course, quite irrational, as time has already shown; Darwin's theory of the origin of species was long ago abandoned.

And Singer's *A Short History of Biology* killed Darwin with kindness in 1931:

> . . . natural selection by the survival of the fittest, is certainly far less emphasized by naturalists now than in the years that immediately followed the appearance of Darwin's book. At the time, however, it was an extremely stimulating suggestion.

In 1981, as the centennial of Darwin's death approached, the staff of the British Museum's Natural History Building in South Kensington unveiled a permanent exhibit entitled "The Origin of Species," a lavish effort in eleven sections, with diagrams, displays, a natural-selection computer game, and a continuously running film loop on which a narrator intones:

> The Survival of the Fittest is an empty phrase; it is a play on words. For this reason, many critics feel that not only is the idea of evolution unscientific, but the idea of natural selection also. . . . The idea of evolution by natural selection is a matter of logic, not science, and it follows that the concept of evolution by natural selection is not, strictly speaking, scientific.

The film loop prompted an indignant editorial in *Nature*: "Darwin's Death in South Kensington." The editorial in turn drew letters from scientists and philosophers in Manchester, Chicago, Brussels, and Odense, Denmark. All that year the journal was full of clashing commentary, for and against: "How True Is the Theory of Evolution?" "Darwin's Survival," "Evolution's Waterloo," "Fit for What?"

Even Darwin himself admitted twinges of doubt. He asks in the *Origin*: "Can we believe that natural selection could produce, on the one hand, an organ of trifling importance, such as the tail of the giraffe, which serves as a fly-flapper, and, on the other hand, an organ so wonderful as the eye?" And though he answers in the affirmative, the question is more than rhetorical, for in a letter to a friend, Darwin confesses:

> I remember well the time when the thought of the eye made me cold all over, but I have got over this stage of the complaint, and now small trifling particulars of structure often make me very uncomfortable. The sight of a feather in a peacock's tail, whenever I gaze at it, makes me sick!

Can the Darwinian process really produce something as marvelous as an eye, a wing, or a feather—let alone a flying bird, a thinking human being? Without being allowed to watch, without the spectacle actually before them, scientists have found it hard to picture how Darwin's process could lead again and again to such magnificent results.

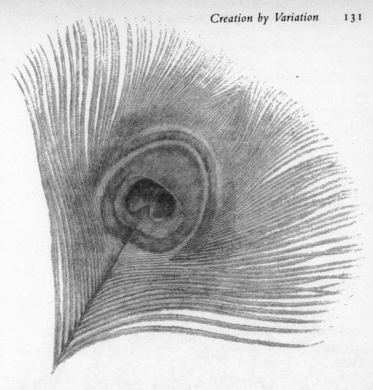

Feather of a peacock. From Charles Darwin, *The Descent of Man, and Selection in Relation to Sex*.   *The Smithsonian Institution*

The mind's eye simply cannot see that far, as the evolutionist George Williams writes:

> I believe that modern opposition, both overt and cryptic, to natural selection, still derives from the same sources that led to the now discredited theories of the nineteenth century. The opposition arises, as Darwin himself observed, not from what reason dictates but from the limits of what the imagination can accept.

Watching natural selection in action is one way to get beyond the debates and abstractions that have wrapped this subject in a century and a half of philosophical fog. The Grants can look and see. And this year, with the help of the hybrids, they hope to see a little more than they have seen before.

A Galápagos finch rides a Galápagos tortoise.
Drawing by Thalia Grant

*   *   *   *

THE STEP FROM INDIVIDUAL VARIATIONS to new species, Peter Grant wrote recently, "will be exercising the minds of evolutionary biologists well into the next century." However it happens, clearly it must happen a lot. Peter often thinks of the way his last teacher, G. Evelyn Hutchinson of Yale, used to put the question: "Why are there so many different kinds of animals?"

In the Galápagos alone there are not only thirteen species of finches, found nowhere else in the world. There are also Galápagos penguins, Galápagos sharks, Galápagos hawks, and Galápagos doves; Galápagos flycatchers, martins, centipedes, butterflies, bees, rats; to say nothing of the celebrated Galápagos mockingbirds and tortoises, and the Galápagos iguanas, Darwin's "imps of darkness."

Why are there so many different kinds of animals? Or plants? There are more than seven hundred different species of plants in the Galápagos, and new ones are still being discovered and described. Almost two hundred of these species are found nowhere else on earth. There is one species of (very variable) Galápagos tomato. There are six species of prickly-pear cactus. There are thirteen species and subspecies of *Scalesia* trees, which, as Peter Grant says, puts them on a par with Darwin's finches. *Scalesia* belongs to the daisy family: a garden-lawn daisy that has grown to the size of a tree.

"The Galápagos seems a perennial source of new things," Darwin wrote to Hooker, after Hooker had sorted through Darwin's Galápagos plant specimens. And of course it is not only the Galápagos that poses the mystery of mysteries. On almost any patch of earth the variety of animals and plants is staggering. Once, as a botanical demonstration, Darwin chose a small fallow field in Kent. It was a poor field without water, its soil "of heavy very bad clay," Darwin writes in *Natural Selection*. On this land in the course of a year a friend of his collected plants belonging to 108 genera. Darwin knew of another botanist "who says that he covered with his hat, (I presume broad-brimmed) near to Lands End six species of Trifolium, a Lotus & Anthyllis; & had the brim been a little wider it would have covered another Lotus & Genista; which would have made ten species of Leguminosae . . . !"

Darwin himself went one better. One February, he relates in the *Origin*, he took "three table-spoonfuls of mud from three different points, beneath water, on the edge of a little pond; this mud when dry weighed only 6¾ ounces; I kept it covered up in my study for six months, pulling up and counting each plant as it grew; the plants were of many kinds, and were altogether 537 in number; and yet the viscid mud was all contained in a breakfast cup!"

All told there are somewhere between two million and thirty million species of animals and plants alive on the planet today. Something like a thousand times that many species—about two billion, by the most conservative guess—have evolved, struggled, flourished, and gone extinct since the first shelly fossils were laid down in the Cambrian explosion, about 540 million years ago. And the great question for the evolutionist is, Why?

Darwin did not claim that natural selection is the exclusive agent in the origin of species. But he did maintain that his mechanism is one way of producing new species, and that it is probably the main way. He stakes this claim in the title of his most important book: *On the Origin of Species by Means of Natural Selection*.

In his first secret notebook he sketched the origin of species as a few rough diverging lines, calling the sketch at first the Coral of Life. One species splits into two, two into four, growing and radiating into branches that will fork again. Later the image was elaborated by Darwin's followers, such as the German biologist and philosopher Ernst Haeckel, into great gnarled oaks with hundreds of species names neatly lettered at the twig tips. That is how David Lack drew his chart of the family tree of Darwin's finches.

Lack's chart represents a close-up of one branch on the tree of life. The species the Grants have chosen for their life's work are half a dozen twigs on this branch. They are the six species that Lack placed in the center of his diagram, the ground finches: three diverging twigs on the left, three diverging twigs on the right.

Darwin's finches, of course, are a classic model of adaptation: generations of textbooks have used their celebrated tool kit of thirteen beaks to illustrate the process. Darwin's finches are also a classic model of speciation: again, they figure in virtually all of the textbooks, very often as the central illustration. That is why these birds have become such a universal symbol of Darwin's process, so that their beaks now represent evolution the way Newton's apple represents gravity, or the apple of Adam and Eve represents original sin.

The standard textbook description of speciation sets the story in the dim past, like a scientific book of Genesis. The textbook diagrams and charts suggest that the flora and fauna of the Galápagos Islands are

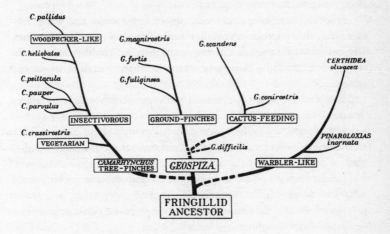

C. pallidus
WOODPECKER-LIKE
C. heliobates
C. psittacula
C. pauper
C. parvulus
INSECTIVOROUS
C. crassirostris
VEGETARIAN
CAMARHYNCHUS TREE-FINCHES
G. magnirostris
G. fortis
G. fuliginosa
GROUND-FINCHES
GEOSPIZA
G. scandens
G. conirostris
CACTUS-FEEDING
G. difficilis
CERTHIDEA olivacea
PINAROLOXIAS inornata
WARBLER-LIKE
FRINGILLID ANCESTOR

From David Lack, *Darwin's Finches*. Courtesy of Cambridge University Press.
Library, the Academy of Natural Sciences of Philadelphia

end points, products of a process of creation that went on once, "in the beginning," and is now more or less complete.

But in Darwin's islands the forces of creation are still at work, in plain view, with "the manufactory still in action." And since Darwin's mechanism for the refinement of a finch beak is also his mechanism for the manufacture of a new species, the Grants' data base allows them to test the power of both processes at once. They can analyze not only the forces that shape adaptations but also, at the same time, the forces that put new beings on this earth.

When Darwin thought about the way the branches of life can grow and split, his mental image was first and always what he imagined had happened in the Galápagos archipelago. Darwin assumed that the summits of these lonely volcanoes were settled by a few chance colonizers that blew or drifted out from the coast of South America. He reasoned that the first immigrants to thrive on the naked lava must have been plants, since any animals that arrived before the vegetation would not have lasted long. A seed-eating finch cannot last a day without seeds.

Conventional wisdom said seeds would be killed by salt water. So Darwin at Down House tried soaking seeds from his garden, including lettuce, carrots, and celery, in little bottles of brine. Then he planted the seeds in glass dishes on the mantelpiece of his study, and was tickled to see them sprout, even after forty-two days in salt water. In forty-two days the average current in the Atlantic could have carried these seeds 1,400 miles. They could have lasted the length of an ocean voyage.

That was the point of Darwin's pottering about with mud from the edge of a pond. "Wading birds, which frequent the muddy edges of ponds, if suddenly flushed, would be . . . most likely to have muddy feet." They could carry mud, and seeds with it, from one place to another.

Darwin was not above gathering bird droppings and picking out undigested seeds with a tweezers. He planted these seeds, and they germinated too. So the first birds to fly over an island could help make the islands fit homes for more birds.

Sometimes his experiments misfired. Darwin complains in one letter that "the fan-tails have picked the feathers out of the Pouters in their Journey home—the fish at the Zoological Gardens after eating seeds would spit them all out again—Seeds will sink in salt-water—all nature is perverse & will not do as I wish it."

But other days his unorthodox experiments left him cackling with glee. He fed oats to sparrows, and fed the sparrows to an eagle and a Snowy Owl at the zoo. Then he waited around a few hours, collected their pellets, and planted them. A seed came up. "The Hawks have behaved like gentlemen," he wrote to Hooker. "Hurrah!"

In the Galápagos, experiments like these are still going on all the time. The archipelago is so young that not even the beginning is over. The very colonization of the islands is in progress, and members of the Finch Unit can watch it unfold before their eyes. When they fly to the islands from the Ecuadoran city of Guayaquil, they look down from their plane windows at the great Guayas River. The Guayas debouches into the convergence of two major ocean currents, the South Equatorial and the Humboldt. Anything rafting down the river can get swept out to sea and westward to the Galápagos. From the plane the Grants often see dozens of natural rafts on the river. The muddy water twists and turns, and long green strings and strands of matted greenery go drifting along with it. Anyone who takes off from or lands in Simón Bolívar International Airport, in Guayaquil, can look down at these fateful rafts: green mats in the dark green delta, perpetually floating out to sea.

In Darwin's view, this was the first step in the origin of Galápagos species—colonization, immigration—and the Grants can watch even this step in action. During their first years in the islands they were surprised to discover half a dozen flame trees growing on the far-northern island of Wolf. Flame trees are common on parts of Santa Cruz and on Santiago, but the Grants could not imagine at first how the trees' seeds had gotten all the way from these islands, which are in the center of the archipelago, to Wolf, at the farthest northern edge of the archipelago, where their red flowers look as striking and deliberately placed as a carnation in a black buttonhole. The Grants found another odd outlier, a single flame tree, old, solitary and flamboyant, on the west side of Genovesa, about 400 meters from the sea, the only flame tree on the island.

On Genovesa, red-footed boobies and frigatebirds often pick up bits of brightly colored plastic from the beaches and drop them far inland or along the tops of cliffs. The Grants, walking along the steep cliff of Darwin Bay, above the landing called Prince Philip's Steps, have often found bottle caps, combs, streamers, and gaudy drift plastic of all kinds. The flame trees' seeds are as red as their flowers, and the Grants wondered if the birds might have picked one up from the beach and carried it inland. To test this idea, Peter, Rosemary, and their younger daughter, Thalia, put thirty flame-tree seeds in a jar of salt water. It was the same sort of experiment that Darwin tried at Down, only the Grants were carrying it out right in the Galápagos, in their camp on Genovesa. Whenever they were back in camp from finch watching they stirred the water in the jar. After three days, most of the seeds were still afloat.

Toward the end of the great Niño, the Grants have reported, they found almost one hundred of these seeds strewn on the southern beaches of Genovesa, where there is not a single flame tree. They found even more seeds of the poison apple, which does not grow anywhere on the island. Apparently these seeds floated to Genovesa from another island, swept there by El Niño, "washed down to the sea," the Grants write, "in the torrents and temporary rivers" that flowed so often that year on Santa Cruz and Santiago. "It is even possible they were carried all the way from the South American continent."

On a typical voyage between islands, the Grants, sitting on a bench at the bow of their boat, may see a green sea turtle, a shark, frigatebirds, dolphins, manta rays, whales, and other lively interisland traffic, including rarities like the Hawaiian dark-rumped petrel, elegant white with black-rimmed wings, beautiful and endangered. Young and immature masked boobies will circle the boat, looking closely at the finch watchers each time they cross the bow, and coming closer and closer with each pass to Peter's outstretched hand until he withdraws it, absentmindedly, talking away. The boobies hover and stare at the humans on deck. The birds' brown heads and bodies as sleek as seals— seals with wings. Young boobies are such strong fliers, they do it so well, that they are a pleasure to watch, and they seem to take pleasure in their motion themselves.

So there are endless streams of wanderers between the islands. When Trevor Price was standing watch on Daphne with his field assistant and childhood buddy, Spike Millington, they kept a bird list, just as they did when they were boys growing up outside London. They spotted twenty-two different species of vagrants on Daphne Major. One spring they heard a Hawaiian petrel calling nightly for almost a week in March, and for another week in April. Sometimes just offshore they saw white-vented storm petrels and an immature great blue heron. They also saw peregrine falcons, yellowlegs, a ruddy turnstone, a willet, an offshore flock of about two thousand northern phalaropes, several species of warblers and gulls, and a dark-billed cuckoo.

Even among the finches on the island, the permanent residents, the citizens of Daphne Major, some are conspicuously more restless than others. In their camp at mealtimes the finch watchers sometimes see what looks like a little bug, growing larger and larger, winging over the crater. It is a finch whose territory is on the other side of the volcano, but who has acquired a taste for exotic travel. Meal after meal, there he is again, a born tourist, hopping in the alleys between the *chimbuzos*, cadging another morsel.

Darwin thought his finches were all more or less marooned, each on its own island, but these birds are still wandering and straying across their islands, all across the archipelago. The Rothschild expedition at the turn of the twentieth century spotted a single specimen flying many kilometers out at sea. Price and Millington found a few tree finches that landed on Daphne but did not stay. They also found a few immigrant *fuliginosa*, *fortis*, *magnirostris*, and *scandens*. Every year some of these species are spotted on the island, usually the ones they call "immatures": the young and the restless. Sometimes after a breeding season a hundred or more immatures will land on an island and visit a while, but invariably, Peter says, they all leave or die before the next breeding season.

So every year some of Darwin's finches go island hopping, landing in new places and new corners, with new neighbors. During the great Niño, quite a few finches from other islands visited Daphne Major, including some *magnirostris*. "The immigrants came in late 1982 and started breeding," says Peter. "Five or six decided to stay. And some of their offspring are still on the island. Dispersal is not unusual. Breeding is." The agitations of the super-Niño shoved these strange finches together, and at the same time, the hybrids began their rise.

Chapter 10

The Ever-Turning Sword

> ... and he placed at the east of the garden of Eden
> Cherubims, and a flaming sword which turned every way,
> to keep the way of the tree of life.
>
> —*Genesis 3:24*

In his earliest secret notes, Darwin wonders if the lines of life in the Galápagos might have diverged simply by adapting to the strange new place in which they had alighted. They might have speciated simply by virtue of their isolation, as they altered further and further over time. In those first notes Darwin imagines that enough small changes accumulating over enough geological time would be able to accomplish almost anything.

Darwin implies as much in a famous scribble he made soon after the meeting with Gould over Darwin's finches: "Let a pair be introduced and increase slowly, from many enemies, so as often to intermarry—who will dare say what result. According to this view animals on separate islands, ought to become different if kept long enough apart, with slightly differ[ent] circumstances.—Now Galápagos tortoises, mocking birds, Falkland fox, Chiloe fox.—English and Irish hare—"

Without question this process of adaptation can and does go on. This is the process the Grants have documented in such detail in Darwin's islands. The adjustments they saw the birds make after the drought, and the flood, are the kinds of changes the finches' ancestors would have made when they landed in these islands for the first time. Each year Daphne is a new island, and its finches adapt to it from generation to generation. The work that began when their ancestors first landed on the island is never finished.

But Darwin's finches are not marooned, each species to its own island. On average there are seven or eight species to each island in the archipelago. Beyond that there is the constant traffic of visiting finches. The birds may have diverged—or have begun to diverge—in isolation, but they are not in isolation now. What happens when lines of life that have begun to diverge in isolation meet up again?

Darwin has an answer, and it is one of the most original steps in his argument.

"I CAN REMEMBER the very spot in the road, whilst in my carriage, when to my joy the solution occurred to me," Darwin recalls in his memoirs, "and this was long after I came to Down." By then more than ten years had passed since Darwin had seen the Galápagos. He had already written out a "very brief abstract" of his secret theory, 35 pages in pencil. He had expanded this abstract into a longer sketch of 230 pages and copied out the whole draft in a fair hand. He had kicked innumerable flints. Yet it was not until that moment, riding on the road, that Darwin felt he really understood the branching of the tree of life.

What drives the branches to diverge again and again? "How, then, does the lesser difference between varieties become augmented into the greater difference between species?" Suddenly Darwin perceived that adaptation to isolated specks of land is not the whole answer. He saw a way that natural selection acting on local varieties, "species in the process of formation," in Darwin's view, "incipient species," would drive wedges between these varieties and push them apart, everywhere on earth.

For Darwin it was like the *eureka* in the legend, when Columbus peers at a butterfly perched on an egg, imagines the earth in the palm of his hand, and suddenly perceives the obvious. It was as if Darwin had taken a few giant mental steps back and, craning his head, stared up for the first time at the whole tree of life.

His vision had nothing to do with star-crossed lovers and the meetings and matings of hybirds. It was more like a vision of war. What Darwin realized is that two varieties living side by side are thrown into competition—competition "in a very large sense," like the contest the Grants are now watching among ground finches on

Daphne Major. Individuals in each of two neighboring varieties will usually find themselves going after the same thing as their neighbors, precisely by virtue of their similarities, like two big-beaked ground finches hunting for the same *Tribulus* seed.

In the struggle for existence one variety or species must often squeeze another. Nearest neighbors, closest cousins, will pinch each other very hard, generation after generation. They collide because they are so much alike in equipment, instincts, and needs. Again, like finches scrabbling through the cinders of Daphne Major for the last *Tribulus* seed, the more alike the varieties the more frequently they will find themselves going for the same seed, the same nook, the same niche, at the same time. Competition through kinship.

An exceptional individual will benefit under these conditions. Any escape, however partial, from that oppressive competition will be an enormous release—almost like finding a fresh island. The lucky individual that finds a different seed, or nook, or niche, will fly up and out from beneath the Sisyphean rock of competition. It will tend to flourish and so will its descendants—that is, those that inherit the lucky character that had set it a little apart. Individuals that diverge from the madding crowd will tend to prosper, while the rest will be ground down.

Selection will act in this way on all neighboring varieties, not just on islands but everywhere on the planet, and the effect will be continually to move varieties apart and repel them. Even if they never actually jostle and joust, never once collide physically over a *Tribulus* seed or a nesting site in a crook of cactus, natural selection will gradually magnify their differences.

At last the two varieties will move so far apart that competition will slack off. It will slack off when the two varieties have evolved in new directions: when they have diverged. Natural selection will have led in effect to another adaptation—the mutual adaptation of two neighbors to the pressures of each other's existence. And the result of this sort of adaptation would be forks in the road, partings of the ways, new branches on the tree of life: the pattern now known as an adaptive radiation.

In Darwin's view, then, each line of life is forever reaching for its spot of sun. Competition among slightly divergent forms everywhere on the planet leads continually to new branches, radiating outward in all directions like a compass rose or the arms of a medieval sun. Darwin called this his principle of divergence.

Again, Darwin never actually saw it happen, though he argued that it could happen, had to happen, and had happened. He pointed to his pigeons. Pigeon fanciers enjoy novelty and they often cultivate the extremes. With tumbler pigeons, breeders had spent several centuries "choosing and breeding from birds with longer and longer beaks, or with shorter and shorter beaks," Darwin notes in the *Origin*. As a result the breeders had created two subbreeds out of one, a long-beaked and a short-beaked tumbler.

In Darwin's view, natural selection too would tend to make nature "more & more diversified." Here the motor is not whim, taste, or love of novelty, as with the pigeon fanciers, but something more basic. The great thing is efficiency—or what the economists of Darwin's day were already calling the "division of labor." "It is obvious," Darwin writes in *Natural Selection*, "that more descendants from a carnivorous animal could be supported in any country" if some were adapted "to hunt small prey, & others large prey." Likewise with herbivores, "more could be supported, if some were adapted to feed on tender grass & others on leaves of trees . . . & others on bark, roots, hard seeds or fruit." In other words, as varieties and species ramify they will become better and better consumers of the world around them, like Jack Sprat and his wife, who between them licked the platter clean.

The advantage of divergence "is, in fact, the same as that of the physiological division of labour in the organs of the same individual body," Darwin writes in the *Origin*. There are certain relatively simple animals in which the stomach ingests both food and air and carries out both digestion and respiration. But a stomach specializing in digestion and a pair of lungs specializing in respiration can each do better work.

No wonder Darwin remembered the exact spot in the road for the rest of his life. It is an extraordinary vision. Natural selection literally organizes life. The process of evolution by natural selection works right up the tree, from individuals to varieties, varieties to species, onward and upward, branching and branching, always diverging, helping to create all the myriad life-forms on the planet. Whole lines die out as inevitably as individuals die out, but the result is always something new and alive: the tree grows.

In this view natural selection is even more powerful than Darwin first imagined. It is both beautiful and terrible, an agent of creation and destruction, like the flaming sword at the gates of Eden, "which turned every way, to keep the way of the tree of life."

*   *   *   *

DARWIN'S PRINCIPLE OF DIVERGENCE has fascinated generations of biologists. But the process is not easy to observe and measure
in action, because it erases its own traces. In fact, Darwin's thesis predicts the general absence of competition. In his view, competition
drives neighbors far enough apart that they are no longer locking
horns, or locking beaks.

The Grants, for instance, can see that *fortis* and *magnirostris* on
Daphne Major are competing for *Tribulus* seeds. But how can they
measure competition among the species that are no longer competing,
like the cactus finches and the *fuliginosa*? If the Finch Unit finds the six
species of ground finches living more or less harmoniously now, how
do they know it was competition that drove them into relative har-
mony? Was the agent Darwin's principle of divergence? Where can
they catch the gleam of the ever-turning sword?

Again, this is one of the most contested questions in all of Darwin's theory, and Darwin's finches have been at the center of the controversy ever since David Lack published *Darwin's Finches* in the
middle of this century. The book created so much excitement because
Lack announced that the beaks of the finches were in fact produced by
Darwin's principle of divergence.

All of the low-lying islands in the archipelago, for example, have
either the small ground finch or the sharp-beaked ground finch—
either *fuliginosa* or *difficilis*. The low islands never have both. The taller
islands in the archipelago do hold both species, and there the two species keep more or less apart. *Fuliginosa* tends to gather around the base
of the island, *difficilis* near the top.

In his mind's eye, Lack saw this pattern as the aftermath of a great
war across the whole archipelago, a war without generals or blood-
shed, between the sharp beaks and the small beaks. Apparently, Lack
said, these two species are so alike that whenever they breed side by
side on the same island they are thrown into competition. If the island
has only one niche for them, then either the line of the sharp beaks
or the line of the small beaks is driven to extinction: one species
outcompetes the other. The technical term for this outcome of a battle between two groups of living things is competitive exclusion. But
if the island is tall enough to offer one species a new niche, a path out
of the race, that species can evolve its way out of the competition.
Then it changes in character. The beak bends, melts, morphs, changes
shape, through evolution by natural selection, until that lineage of

birds is freed from the dreadful war. The technical term for this out-come is character displacement.

Lack pointed out that on Santa Cruz the small beaks are small and the medium beaks are medium. On Daphne, where there are very few small beaks, the medium beaks have become smaller. On Los Hermanos, where there are very few medium beaks, the small beaks have become larger. Lack saw this same pattern with many other pairs of finches in the islands: again and again, traces of character displace-ment.

In the 1950s and 1960s, Lack and others explored competition theory and its ramifications. They pursued the theory so triumphantly and unanimously that a few ecologists and evolutionists with other in-terests began to feel competitively excluded. One of them was an American ornithologist, Robert Bowman, who went to the Galápagos in 1952, after Lack, and studied the Galápagos finches as his thesis project. This was before the invention of mist nets. To find out what the finches were eating, Bowman shot them by the hundreds and in-spected the contents of their stomachs. He decided that what a Galápagos finch eats does depend on its beak—a conclusion that would later be confirmed and extended in such extraordinary detail by the Grants' International Finch Investigation Unit.

However, Bowman also noticed, more than Darwin or Lack had noticed before him, that Galápagos plants, like Galápagos finches, vary from island to island. Bowman wondered if this variation in itself might explain the branching of the finches' lines. The birds might have evolved their distinctive beaks and habits by adapting to the local varieties of flowers and seeds as they settled on island after island. They might have speciated, in other words, according to Darwin's first vi-sion of the evolutionary process, before his *eureka* in the carriage. That was a simpler story, Bowman argued—and who could prove it wrong?

Time and again Bowman watched the sort of mixed flocks that Darwin himself had noticed when he first got off the *Beagle*. Time and again he saw four or more species of ground finches feeding in the very same bush, small peaceable kingdoms of ground finches. "And since there is no direct evidence that competition is occurring at the present time," Bowman declared in his thesis, "I see no logical reason to assume that it must have occurred in the past."

To argue that the absence of competition is a proof of its power seemed to Bowman infuriatingly circular reasoning. There were other ecologists who agreed with him. A pitched battle began between the competitive camp and the noncompetitive camp, and in the absence of hard evidence and actual observations it dragged on and on.

The ecologist Joseph Connell declared in a celebrated manifesto that until someone came along with harder facts, he would no longer buy the old guard's arguments for Darwin's principle of divergence: "I will no longer be persuaded by such invoking of 'the Ghost of Competition Past.' "

The ecologist Daniel Simberloff was even more emphatic. He argued that Lack's famous patterns might be nothing more than the faces we see in the moon, in clouds, in Rorschach inkblots. Even if we see species living together in different niches, that does not prove that they drove each other into those niches—that they repelled each other like the identical poles of two magnets. They might instead have diverged by random adaptations toward different seeds and different needs. Finches colonizing the islands at random would have produced patterns that could be interpreted in all kinds of ways. The finches' distributions in the archipelago might be "most fruitfully treated as a random process."

Peter Grant suspected that Lack was right; but he too was annoyed by Lack's dogmatic style, his willingness to issue pronouncements without having seen what is actually going on. Grant once accused Lack in print of a "distressing lack of objectivity." . . . . "Future generations," he wrote, "will surely wonder that we discuss competition so much yet know it so little and measure it even less."

THAT WAS ONE OF THE FASCINATIONS of the events the Grants witnessed in their first dry season, events they still see repeated each year. Again and again, when the rains end, their flocks of Darwin's finches move apart and specialize, *magnirostris* going this way, *fortis* that way, *scandens* and *fuliginosa* going off on their own as well, each according to its beak. Not only does this divergence help to prove that the beaks of the finches are adaptive; it also helps to prove that their differences have survival value. It suggests that the forces Darwin envisioned in his carriage more than a century ago are at work right now among his finches.

When birds that have been eating the same foods for months begin to fan out into specialties, according to the sizes and shapes of their beaks, we are seeing a kind of divergence in the course of a sin-

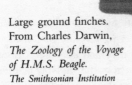

Large ground finches.
From Charles Darwin,
*The Zoology of the Voyage
of H.M.S. Beagle.*
The Smithsonian Institution

gle season. And we can also see that the consequence is just what Darwin imagined: for as each dry season wears on and the birds are scrabbling for food, their choices as they hunt and peck diverge more and more, which reduces the competition between them. That they respond to the drought in this way, in effect making room for one another, may help to explain why they are able to coexist on the little island at all.

Of course in the dry season the birds are only changing their behavior, not their beaks. Their diets diverge as their stocks of food are depleted because so many individual birds stop trying to do everything and concentrate on what they do best, according to the beaks and bodies they were born with. Changing your behavior is much quicker than changing your anatomy. But the change is driven by the same sword of competition. Darwin's finches are responding to the same forces that Darwin believed led to their creation.

The traditional picture of the economy of nature, the picture most of us learned in school, is that nature is a stiff hierarchy, a food pyramid of more-or-less rigid trophic levels. Plants are at the bottom of the pyramid, making food from sunlight. Herbivores feed on the plants, and carnivores feed on the herbivores. Within each level there are assorted specialists, called guilds; any ecologist can rattle off long lists of them: "leaf eaters, stem borers, root chewers, nectar sippers, bud nippers"—much like the feudal guilds of shoemakers, tailors, butchers, bakers, and candlestick makers.

As more and more ecologists and evolutionists watch life up close and long term, the way the Grants are observing Daphne Major, they see that these categories are not as fixed as they had imagined. More and more naturalists are shifting their efforts from the study of pattern and structure to the study of process and motion, watching change through time, and, like the Finch Unit in the Galápagos, what they see again and again is the lesson of Heraclitus: everything flows. Nature is fluid. The finches on Daphne start out their year in a single guild, the guild of the ground finches; then in hard times they break up into smaller guilds. Other ecologists and evolutionists are watching this same process of splitting and shifting guilds elsewhere. This fluidity of nature suggests that as divergent pressures are repeated year after year, animals and plants around the world are even now evolving differences in physique and taste that (if they continue) will make their guilds diverge farther and farther.

Of all the watchers who have worked with the Grants in the

Galápagos, the one who has grown most fascinated with Darwinian divergence is Dolph Schluter. Schluter is a rising star now in the field of evolutionary biology, but he was about to drop out of school when he met the Grants. He had just completed an undergraduate degree in wildlife management at the University of Guelph, in Canada. He hadn't applied to graduate school, even though he had the highest marks in his class. He was sick of school, burned out. He was planning to take a job trapping muskrat and mink in the Athabasca Tar Sands, in Alberta. Then he heard a student of the Grants give a seminar.

"I'd never realized people *did* this work," Dolph says now. "I knew about evolution, of course. But it had never struck me that someone could actually study it—in modern times. My idea somehow was poring through fossils. That someone could watch evolution! It was a revelation.

"I wrote to Peter Grant. It was the only place I applied. Otherwise I'd still be trapping muskrats."

Dolph's Ph.D. project under Peter Grant was a close-up study of the war between the small beaks and the sharp beaks on a single Galápagos island. Dolph read the latest manifestos of Simberloff et al., which called finch distributions "a random process," and he went down to the islands in a state of radical doubt. "I was quite dizzy with the idea that everyone—including Grant—had overestimated the importance of competition."

The whole Grant family rode the boat with Dolph, escorting him to the island of Pinta, one of the most remote in the archipelago. "As Pinta came into sight—and it was spectacular! I'd never been south before, let alone the Galápagos; a school of dolphins was leaping in the boat's bow wave—Peter Grant turned to me and said, 'Well, Dolph, you know, there might only be a master's in this.' Only a master's thesis, not a Ph.D.

"We had no idea what we'd find, in other words," Dolph explains, with laughter in his voice. "Grant was trying to say the idea might not be feasible. The plans might not work."

The Grants taught Dolph how to set up mist nets, how to catch a bird and measure it. They stayed a week. "Then it was, 'Goodbye, we're off to Genovesa. See you in five months.'" Schluter and his field assistant were now the only human beings on the island.

Like all the taller islands in the Galápagos, Pinta has an amazing slope. At the bottom, it is a desert of bare black pahoehoe lava and thorny scrub: dry, hot sheets of bare rock. As you march up the slope you walk into a cloud, and a green forest canopy begins to close over your head. The highlands are moist and cool; the ground is thick and soft underfoot with deep forest litter. At the top is elfin forest: misty, green, with moss-covered trees, lichens, orchids.

As soon as he had learned to tell them apart, Dolph could see that the small beaks' and the sharp beaks' territories overlapped. *Fuliginosa* went up to the summit of the island, and *difficilis* went not quite all the way to the bottom. "I saw that right at the beginning," he says. He had expected to find the two rivals living in separate zones of the island, bottom and top, as Lack had described, with territorial skirmishes all along the border. "That possibility was completely shot," Dolph says. All was quiet on the western, eastern, southern, and northern fronts. *Fuliginosa* and *difficilis* were ignoring each other. "This is typical of fieldwork," he says now. "You plan and plan, and then you get there and realize you can't do it."

In the next few months, he and his assistant set up study grids up and down the slope. They found to their astonishment that the finches, although they overlap in territory, overlap hardly at all in diet. On Pinta *fuliginosa* takes seeds out in the open, pecking on the hard hot lava, mostly down by the shore. But *difficilis* scratches with its long claws in the leaf litter, mostly up in the forest, kicking the leaves away, shoving pebbles aside, and snatching up spiders and snails, crickets and caterpillars with its beak. *Fuliginosa* eats hardly anything but seeds; *difficilis* takes hardly any seeds. So if Darwin and Lack were right about the cause of the divergence, there was no way to tell by watching these birds. "There's very little trace of competition," Dolph says cheerfully. "It really is . . . a ghost."

Because he could not settle the competition question on that one island, he went on to another island, and another. Over the next few years he mapped Lack's patterns in ever-finer detail from one end of the archipelago to the other. The masses of measurements that he and others collected were consistent with Lack's predictions, consistent to a remarkably detailed degree.

Take *fortis*, the medium-sized ground finch. Its equipment represents a trade-off between the best beak for small soft seeds and the best beak for big hard seeds. The bird evolves toward the best deal it can make, and on each island the best beak is determined partly by the presence of *fuliginosa*, because small beaks in the dry season will lower the pile of small soft seeds. A *fortis* with a beak too small to crack a *Tribulus* will have to compete with *fuliginosa*, but *fuliginosa* are more efficient than *fortis* can be. So the smaller *fortis* will die younger than the bigger *fortis*. In this way competition with the small beaks drives up the average size of the medium beaks. On islands where it has been driven up in size by this competition with small beaks, it is much better with *Tribulus*, and it eats fewer small seeds. *Fortis* on Daphne has very little competition from *fuliginosa*, and there *fortis* throws down most of the *Tribulus* it picks up. On the other hand, *fortis* on Pinta, Marchena, and Santiago, where there are many *fuliginosa*, eat fewer small seeds but seldom meet a big seed they cannot crack.

Meanwhile, where *fuliginosa* is competing with *fortis* it tends to specialize in the smaller seeds because *fortis* is so much more efficient with the big hard seeds of *Tribulus* and the prickly pear. But where it has an island all to itself, as on Los Hermanos, the beak of *fuliginosa* is big and deep, and it can handle fairly big seeds. On Los Hermanos its beak is almost identical to the beak of *fortis* on Daphne.

On the islands where *magnirostris* is missing, the beak of *fortis* is bigger than average; on the islands where *fuliginosa* is missing, *fortis* is smaller than average. It fills both its own niche and that of its missing rival.

There are far too many pairs of species like the sharp beaks and the small beaks, or the small beaks and the medium beaks, or the medium and the large, that have divvied up the islands in this neat way between them, converging on the same beak where they live apart but diverging where they are neighbors, making an infinity of finely calculated mutual adjustments, for the patterns to be due to chance. Lack was essentially right. "He was not right in every detail," Dolph says, "but the mechanism was dead-on. I've since concluded that he was a pretty bright fellow. He's my hero actually. He had a disarming writing style, and he could make very forceful arguments. In fact, I think they should be called Lack's finches and not Darwin's. Darwin didn't see the significance of the birds. He thought there was just one species per island. He didn't even try to pull it together—he didn't do a bloody thing with them except collect them. That's why they're Lack's finches and not Darwin's."

AFTER A FEW YEARS in the islands, Dolph arrived at a new way of visualizing the Darwinian pressures on Darwin's finches.

Imagine a single population of birds on a desert island. As this population changes through time, generation after generation, it is perpetually moving toward the optimum design for that island. It is working toward its point of maximum fitness, toward what evolutionists sometimes call an adaptive peak.

Picture that point of maximum fitness, the peak of optimal design, as the summit of a mountain. The slopes around the summit represent designs that are slightly inferior to the design right at the peak. The valleys far below represent designs that are fatally inferior to the design at the peak. A winner, a bird with the right beak, the right wings, the right tarsi, is on the peak; a loser is wandering down in the valley.

If a bird is blown out to sea and lands on a desert island, it will not be perfectly adapted to its new home from the start. Its fitness will be somewhere down on the slope of the adaptive landscape. If it is very far down in the valley, it will die, and its line will die out. But if it can hang on and breed, its descendants will move up and up, evolving and adapting, until they reach the peak.

Evolutionists have been thinking about adaptive landscapes for a long time (the idea was introduced by Sewall Wright, one of this century's great Neo-Darwinians, in 1932). The adaptive landscape is one of their favorite metaphors for evolutionary change: almost as popular as the tree of life itself. But Dolph was the first to see how revealing the idea could be when applied to Darwin's finches. Peter Grant had asked him to find a way of pulling together all of the group's work on beaks and seeds in one framework; to create an all-embracing mathematical model that would help the finch watchers understand what they were seeing. Dolph decided to draw their data in terms of adaptive landscapes, with the aid of computers. Dolph is one of the most sophisticated mathematicians in the Finch Unit, along with Trevor Price, his old field mate and office mate, with and against whom he has honed his skills.

In picturing the forces that lead to character divergence, the adaptive landscape makes a better metaphor than the tree of life. It permits a finer level of description. A population of finches splitting in two (the forking of a branch, in the tree-of-life metaphor) can be imagined as setting off on a sort of lonely pilgrimage or mass migration across an adaptive landscape. Some birds leave one peak and migrate to another. First there is just one species in the landscape, clustered around one peak. Then there are two, clustered around two neighboring peaks. In between there lies a valley.

It was in 1979 that Dolph realized how he could use this image of the adaptive landscape to test Darwin's principle of divergence with

Darwin's finches. He realized that he and the other members of the Finch Unit had amassed so many measurements that he could now draw a fairly realistic adaptive landscape for the Galápagos ground finches.

The unit had measured all of the seeds these ground finches can eat, and many of the seeds they can't or won't eat. Dolph knew what kinds of seeds are found on each island, and in what season, and in what amounts. He entered that information into a computer.

Dolph also knew how big a beak can crack how big a seed. The critical dimension here is beak depth: the deeper the beak, the bigger and harder the seed it can crack. He entered that relationship into the computer.

Finally, he knew how many seeds it takes to keep a finch alive. He entered that relationship into the computer too: such-and-such a mass of seeds translates into such-and-such a mass of finches.

"*Now,*" Dolph says, in the tone of a geometrician who is trying to arrive at first principles: to use just a few lines and points and the minimum of rules, and from that generate a globe, a world, a universe. Suppose a single species of finch lands on a single Galápagos island. Suppose it meets no competition, and it survives and multiplies. What size beak will it evolve?

Dolph told the computer the range of beak sizes available to all of the ground finches on the islands, from the smallest to the largest ever measured. He also told the computer the range of seed sizes available on the islands, from the smallest to the largest, and how many of each size there are. Then he programmed the computer to calculate how many finches a hypothetical island could support, given finches with beaks of every possible depth, from the shallowest to the deepest finch beak on record anywhere in the world, at increments of a small fraction of a millimeter. He programmed the computer to make the calculation over and over.

Dolph assumed that it would draw a peak corresponding to the best of all possible beaks, that is, the beak size that would produce the maximum number of finches on that idealized Galápagos island. The peak the computer drew would represent the best possible beak on the island. The valleys on either side of the peak would represent all the miscellaneous beak sizes that were relatively unfit.

He set the computer going, and he was thrilled by what he saw. The computer did not draw a single peak. Instead it drew three peaks, with deep valleys between them.

The computer had picked up something that is not obvious to the eye. Seeds in the Galápagos come in three types, for a finch: easy, medium, and hard. Grass seeds are among the smallest and easiest, *Tribulus* seeds are among the largest and hardest, the seeds of the passionflower are among those in between. There is a continuous range, but on the whole the seeds of the islands do cluster into these distinct groups. Accordingly, the computer had drawn three peaks, for three beaks, one above each of the three piles of seeds.

In effect the computer had predicted three species of finches, each with a beak precisely the size to crunch the seeds in its pile. Three peaks meant there could be not one but three highly adaptive beaks for ground finches on that idealized island. There was more than one good beak. A single founding species alighting in that adaptive landscape could diverge into three.

Taking advantage of the Finch Unit's masses of data, Dolph programmed the computer to run a more realistic simulation. He asked it to calculate the beaks that should have evolved on a dozen islands of the Galápagos, given the seeds that are actually found on each island.

With just three factors, beak size, seed size, and competition, the computer predicted correctly the divergent paths of evolution for the beaks of finches on every one of the islands.

Once again Darwin's finches are true to their destiny as paragons of Darwin's theory. They are the best case of character displacement ever found.

"THE STRUGGLE," Darwin writes in the *Origin*, "will generally be more severe between species of the same genus, when they come into competition with each other, than between species of distinct genera." Yet in principle Darwin's process of character displacement can occur even between living things spaced far apart on the evolutionary tree, if they are thrown into competition. There can be competition between kingdoms: say between plants and animals, or plants and insects, or insects and bacteria. Two British evolutionists, Michael Hochberg and John Lawton, have made a calculation that they find "both instructive and astonishing." Suppose there are three hundred thousand species of higher plants. Higher plants include ferns, conifers, flowering plants—essentially anything green that grows more than a few inches off the ground—and three hundred thousand species is a conservative estimate. If every one of those species is fed upon by just ten species of insects (also a conservative estimate), and if each of those insects is the host for one infectious disease and five species of parasitoids (also conservative), then there is room for about fifteen million different competitive interactions at this single busy intersection in the kingdoms of life.

Almost no one has looked for divergence in action on this front, but Dolph Schluter has. He suspects that distant branches in the trees of life may be competing right now in the Galápagos.

Some of Darwin's finches drink flower nectar, which finches do not do on continents. They can get away with it in the islands because they have little competition for it: not many pollinating insects have crossed the Pacific and colonized the archipelago. Only one species of bee has made the journey so far, a large carpenter bee, named for the naturalist who first collected it: *Xylocopa darwini*. Darwin's bees have not made their way to the northernmost islands, but they are found in most of the southern islands. There they compete for flowers with the finches.

The finches on the northern islands get most of their nectar from the little yellow flowers of *Waltheria ovata*. In the dry season, when food is hard to find, small beaks and some sharp beaks will drink nectar from *Waltheria*. The sharp beak inserts its whole long beak to probe the flower, while the small beak sticks in only its lower mandible. Both of these finches have gotten very good at the work: they can probe as many as forty flowers a minute.

On northern islands like Genovesa, where there are no bees, nectar makes up about 20 percent of these finches' diet; it is an important dietary supplement to the small seeds they spend their lives scrabbling for. But on southern islands, where there are bees, nectar makes up less than 5 percent of the birds' diet.

Dolph has noticed that the nectar-drinking finches tend to be smaller than finches of the same species on islands where there are bees. For instance, the average wingspan of small beaks is almost 5 millimeters shorter on the islands of Pinta and Marchena (no bees) than on the islands of Fernandina, Santa Cruz, Santiago, Española, and Isabela (bees).

Dolph once spent two weeks observing a spot on Marchena that was rich with yellow *Waltheria* flowers. He saw finches fighting to defend clumps of these flowers. Small beaks that owned a few clumps would defend them against other small beaks. Dolph netted and measured as many of these finches as he could. The average nectar drinker was significantly smaller, and weighed about a gram less than the average teetotaler. That is, even on Marchena, only the finches that were cut out for flower feeding by their small size were drinking nectar.

Apparently, then, on islands where there are no bees, selection pressure has reduced the size of two species of Darwin's finches, and these birds have squeezed into the bees' niche. On islands where bees have arrived, the birds have been forced back out of the niche.

If so, this is a case of character displacement not just between sibling species but between a vertebrate and an invertebrate, which goes far beyond what Darwin imagined. We never see them fight over a flower, but there is no peace between the birds and the bees.

IS CHARACTER DIVERGENCE UNIVERSAL and powerful, as Darwin thought, or is it rare and weak, as some evolutionists argue today? "My own guess is, character divergence is likely to be quite common and important," says Peter Grant, "but I think of rather small magnitude, in terms of the quantity of the evolutionary shift."

"You mean . . . ?" asks Rosemary.

"Take two species that come together. Say they differ by 10 percent. They would need to be about 15 percent different to coexist without serious competition. There are two possible outcomes. One of the two may go extinct, or the two species may diverge until they are 15 percent different. That shift is not very great: only another 5 percent."

"I see. Okay," says Rosemary.

"Now if you agree with that, how easy would it be to see the evolution of that 5 percent shift? The answer is, not at all easy. You'd need a lot of detail. Character divergence could be very common but simply not be known."

In Peter's view, most of the divergence among Darwin's finches takes place when they are isolated, living on separate islands. Then, when they come together and share an island, competition drives their lines farther apart. This divergence is a simple consequence of the struggle for existence, just as Darwin imagined in his carriage. It pushes the finches one or two steps closer to the origin of species.

Chapter 11

Invisible Coasts

Cleopatra's nose, had it been shorter, the whole face of the world would have been changed.

— BLAISE PASCAL,
Pensées

When Darwin was hobnobbing with pigeon-men it was the power of selection he wanted to hear about, not the power of crossing. "I sat one evening in a gin palace in the Borough amongst a set of pigeon fanciers," he writes gleefully to a friend, "when it was hinted that Mr. Bult had crossed his Pouters with [larger] Runts to gain size; and if you had seen the solemn, the mysterious, and awful shakes of the head which all the fanciers gave at this scandalous proceeding, you would have recognized how little crossing has had to do with improving breeds."

"It has often been loosely said that all our races of dogs have been produced by the crossing of a few aboriginal species," he writes in the *Origin;* "but by crossing we can only get forms in some degree intermediate between their parents. Did the Italian greyhound, bloodhound, bulldog, &c [exist] in the wild state?" No, of course not. Ergo, he writes, "the possibility of making distinct races by crossing has been greatly exaggerated."

Darwin was so eager to build a case for natural selection that he may have missed something here. Hybrids may have been more important to most breeders than he thought.

At the turn of the century, the American geneticist Raymond Pearl, adopting a so-show-me attitude, cultivated and quizzed hundreds of breeders, just as Darwin had done. Pearl asked bantam breeders, for instance, if they could document a case in which a new bantam line was "created solely by selection of small-sized individuals of a large race, variety or breed of fowls, without any crossing in of bantam blood?"

Pearl got his answer from the world's leading authority on bantams, one J. F. Entwisle. "If such can be done," Entwisle wrote, "then our thirty-odd years' experience of bantam 'manufacturing' counts for very little. We have lived to see the manufacture of some forty varieties, and none without crossing, so far."

"Darwinian selection plays an extremely minor and unimportant part in the process as it is actually performed," Pearl concluded, swinging the other way from Darwin.

Today, unlike Pearl, the Grants know that Darwinian selection does work. They have seen it work. Now, as they contemplate the rise of the hybrid finches, they are beginning to suspect that selection and crossing work together as part of the same creative process.

THERE ARE TWO KINDS of shorelines in the islands, the visible and the invisible.

The visible shores are the black broken rocks and white broken waves where volcanoes rise out of the Pacific. They are the borders of the air, the sea, and the lava, wave-gnawed rings around the summits

The Kicker Rock. From Charles Darwin, *Geological Observations.*
The Smithsonian Institution

that have given Darwin's finches their homes in the middle of no-where. These shores are defined simply by the level of the sea.

The invisible shores are the borders between the birds themselves. These shores are more intricate. They are defined by the secret codes and unwritten rules that wrap each of the thirteen Galápagos finches in a kind of self-invented isolation. These boundaries hold each species apart from the rest (with the interesting exception of the hybrids), so that even though seven or eight of them may share the same volcanic summit, feed together in mixed flocks, scrape the same cinders for the same seeds, they still breed as much apart as if they were themselves an archipelago of enchanted islands.

Darwin helped to map the visible coasts of rock and coral sand—the *Beagle* was a surveying ship. But his thoughts about the invisible borders between species were inconsistent and confused. In his first se-cret notebooks he described species as isolated by their sexual instincts and equipment. Later he became so absorbed in his principle of diver-gence that he neglected this aspect of his subject. What holds the new species apart? What are the barriers, and what makes the barriers harder or easier to cross? These are coasts that Darwin left more or less unsurveyed and unexplored.

Evolutionists understand now that the isolation of species is not merely a matter of populations locked apart by mountains, canyons, or seas, as Darwin's finches are held loosely apart by the visible shores of the Galápagos. The isolation of species consists chiefly in the invisible barriers that can carve off a corner of a population into a new island or carve a single large population into a set of scattered, more or less lonely gene pools.

It is not just intersterility that produces these invisible shorelines, as between horses and donkeys. Nor is it physical or physiological in-compatibility, as between elephants and their fleas. Darwin's finches can interbreed and produce fertile young, but something keeps most of them from doing it.

The barriers around the birds are invisible because they are created by the creatures' own behavior. It is not anatomy but instinct that holds them apart. To inquire into the origin of species is to inquire into the formation of these instinctive, invisible barriers. Though Dar-win himself had trouble with the subject, the secret of the origin of species lies somewhere along these viewless coasts, these shifting shorelines among living things.

Clearly, once varieties begin to split off from one another, they need some way to keep themselves apart. Whether they diverge in isolation, or partway in isolation and partway in each other's company, at some point they do have to learn to breed with their own kind—their own *new* kind. Otherwise a new line would blend with the lines around it and disappear. Darwin was impressed, for instance, by the work it took to maintain fancy lines of pigeons. If breeders crossed their birds carelessly their creations would soon revert to their starting point, the common rock pigeon. All the pigeon fanciers' whimsical creations, all those pouters, fantails, laughers, and scandaroons, would be lost.

What holds these breeds apart and prevents their collapse to the origin is the pigeon fanciers' continual vigilance in arranging crosses. The best of the breeders in Darwin's day, like the famous pigeon-man Sir John Sebright, could tell by eye a difference in beak of one sixteenth of an inch. Darwin writes (pointedly) that he had to work hard to persuade these exquisitely sensitive pigeon fanciers to make the kinds of crosses that would undo their lines. Such experiments went against the grain, he says—even in the name of science, even for a fee. Darwin had to cajole and wheedle, reminding the breeders that after all, even ugly birds are "still good for the pot."

It is Darwin's argument that natural selection can arrange matches too, and better than the best human breeders. The selection process will give living things this gift, this talent, simply by choosing the fittest variants in the nest, the litter, and the seedbed. It will improve their ability to tell one another apart as naturally and inevitably as it improves their ability to eat or to fly. Selection will act this way because those individuals that make bad choices in their mates will suffer a disadvantage: their offspring will be less likely to prosper in the struggle for existence. Those individuals that make good choices will enjoy a corresponding advantage: their offspring will prosper.

That is why hybrid finches should be rare on Daphne Major. In a population that is diversifying, evolving distinctive adaptations, selection will tend to preserve a new and valuable invention because individuals in a gradually diverging line will have an adaptive advantage if they choose, as mates, others in the new line. If they choose a mate within the line, they stand to perpetuate the new adaptation. If they do not, they stand to lose the adaptation: they lose the family jewels. So there is selection pressure on creatures like Darwin's finches to learn to discriminate. Those birds that evolve a sexual taste for the new line will do better than those that retain a taste for the old line.

In this way, natural selection will ensure that the birds' own powers of discrimination are finer than the pigeon fanciers'. It will make birds in the wild, as Darwin put it, "their own Sir John Sebright." The power of natural selection will make birds of different lines sexually repugnant at mating time.

Darwin saw some evidence of this kind of mutual repugnance even in the pigeon coops. In *Natural Selection* he reports that "the Dovecot pigeon, the ancestor of all the breeds, seems to have an actual aversion to the several fancy breeds."

But when Darwin describes the finches of the Galápagos ("probably the earliest colonists, having undergone far more change than the other species"), it is the visible, physical isolation of their islands that he stresses, not the finches' invisible, instinctive isolation—which Darwin knew nothing about.

AMONG DARWIN'S FINCHES, a mating season begins with all the black cocks on the island singing in the rain. Each male broadcasts from his singing post, and while he sings he scans his territory. If a female alights near one of the display nests that he has built, he darts from his singing post and flies to her. If she is one of his own kind, he sings and shakes his wings at her, makes them quiver tremulously. Then he flies to the nearest nest. (Any nest will do, even the nest of a rival if the rival is out.) He goes in and out, in and out of the nest, looking back at the female over his shoulder. Sometimes he picks up a bit of grass in his beak and puts it down again, rapidly, over and over, as if he were trying to catch her eye.

If the female hops toward him, or at least if she does not hop away, the male begins what the Grant team calls the "sex chase," flying after her on a "twisting and undulating" flight path. He swoops and pivots in mid-air in what looks to the finch watchers like maneuvers to keep the female on his territory. The female stays just a wing-beat ahead of him making high-pitched cries, *"Kew, kew, kew!"*

These are ground finches, remember. Their wings are short and stubby; they are not really cut out for this kind of aerial acrobatics. The long twisting flights must require fantastic amounts of energy to keep up, especially after a long dry season of thirst and famine. In other words, the sex chase is expensive. But all the finches do it when they court, young and old alike, even birds that have mated together for years.

Laurene Ratcliffe and Peter Grant watched more than a thousand finches pair off and go chasing each other through the air. In all those courtships, they saw a male go after a female of the wrong species only twenty-six times. And in the middle of each of those rare cases, either the male broke off the courtship or the female flew away. In four of the mismatched courtships, the male stopped flirting and actually attacked the female.

Long ago, before Lack, an ornithologist speculated that Darwin's finches might all be one intermingling species, a hybrid swarm: finches just mated with finches. That speculation (Peter Grant calls it a "cry of desperation") could only have come from a scientist who had studied the finches in the museum trays and not out on the islands.

Since the very first year of their watch, the Grants have known that Darwin's finches have a talent for telling one another apart. The birds' ability to recognize one another is impressive, even a bit mysterious, given the remarkable range of variation in their beaks and bodies and the drab sameness of their plumage. It is true that the six species of ground finches look different from the six species of tree finches. The tree finches' plumage is a yellowish-green, the general tint of Galápagos trees in leaf, and the ground finches' plumage is black and mottled brown, the general tint of Galápagos lava. They are markedly different birds, even though the tree finches do spend much of their time on the ground and the ground finches spend much of their time in trees.

But there is no obvious difference in plumage among the six species of ground finches. There is no obvious difference in their styles of courtship either; the singing post, the flirtations at the display nest, the sex chase—all seem to be the same from species to species. There is no great difference in their territories either. On Daphne Major in good years a map of the male finches' turf is a sprawling jumble of intersecting circles and oblongs. A male finch won't let another male finch near its nest, so the epicenters of these territories, their capitals, so to speak, are always separate. And a male *scandens* will not let another male *scandens* claim his territory, so their turf is always separate. But the territories of *scandens* will overlap with the territories of *fortis*. On a map all these territorial borders look like the ripples of raindrops in a pond. In the middle of this overlapping confusion, every *fortis* has to find a *fortis;* every *scandens* has to find a *scandens.*

The advertising song that each male broadcasts from his singing post, or from his nest, is distinctive enough that the Grants and the other finch watchers learn to recognize some individuals by their songs. And this is something birds can do too. Male hooded warblers can tell their neighbors apart by their songs during the breeding season. Even after an eight-month hiatus, during which the warblers stop singing, migrate to Central America, and then return to their breeding territories, the males still remember and recognize their rivals' songs.

Among Darwin's finches there are usually not one but two or more versions of the advertising song in circulation within each species, on each island. Some males sing one version, and some males sing another. On Daphne and Genovesa, a very few birds sing more than one. But the average male has a very limited repertoire. He has just one advertisement for himself, which he sings over and over all his life.

The Grants have spent many hours listening to the songs in the islands, and studying the song-graphs at Princeton. In the graph, the song of one of Daphne's champion breeders, Number 2666, is a sort of ghostly pyramid of six gray, fine-lined blotches. Finches Number 2663, 2664, and 2665, the siblings of 2666, sang exactly the same song, and so did their father. In fact most males sing the song of their father. The Grants have recorded the song of 2666 year after year, and the song-graphs are all the same, they don't change. "Peter says it sounds like '*Mostly Muesli*,'" says Rosemary, and laughs. "It goes *Mostly Muesli, Mostly Muesli, Mostly Muesli*. . . .'"

Another advertisement song on Daphne is more of a *chuh-chuh-chuh*, she says. *Scandens* is mainly *chuh-chuh-chuh-chuh-chuh*. Rosemary can sing them all aloud as she leafs through the song-graphs. "*Chae-ae-ae*. This one is *wicki-picki, wicki-picki, wicki-picki*, very fast."

Besides their advertising songs, all of the finches can also produce a sort of hissing whistle, a very high, descending note. All of the ground finches' whistles sound much alike (although on the island of Pinta the sharp beaks add a distinctive, drawn-out *buzzz-clink!*). So it is the advertising song that the Grants have studied most closely as a clue to how it keeps the species apart.

Laurene Ratcliffe has discovered subtle variations even in songs that sound alike in the islands and look much alike in the song-graphs. These variations are picked up and passed down from father to son. In a series of experiments, she and Peter Grant have taken tapes of some of these songs and broadcast them from a speaker in the center of a finch's territory. More than nine out of ten males reacted strongly and aggressively to the broadcasts. They seem to respond equally to either of their own species's song types, A or B. That is, whether they themselves sing song A or song B, they will fly over the speaker, countersing with it, land on it or stand in front of it, and stare at it from close range. They react as if to drive away a strange male who has set up shop and begun to sing in the middle of their territory. Even on islands where two species sing much the same song, a male will respond much more vigorously when it hears one of the songs of its own species coming out of the loudspeaker than when it hears a song of the other species. Laurene and Peter tried these kinds of tests with several different pairs of species of ground finches on several islands, always with the same results.

Singing the right song is important; a finch that learns the wrong song is in trouble. The finch watchers have seen, year after year, a cactus finch singing away from his tallest cactus in the prime of health. Year after year he sings the song of *fortis*. He is not getting any younger, a good bird and a good singer doomed to a genetic death.

Apparently, every once in a while, instead of learning the song of his own father, a male finch learns the song of a neighbor; once in a blue moon, he learns the song of a neighbor of the wrong species. None of the finch watchers has ever seen this kind of accident in progress, only the result, full-grown ground finches singing the wrong song. Perhaps, Peter Grant writes, "they lost contact, as fledglings, with their fathers, were fed by a male of another species, and became misimprinted on his song." They may even have acquired the strange song in the nest, Peter believes. There are opportunities. Dolph Schluter says he once had a nest on Pinta in which a female sharp beak and a female small beak each laid three eggs. All six eggs hatched, and both pairs of parents, the sharp beaks and the small beaks, squabbled over ownership of the nest until the sharp beaks won out. Unfortunately the six chicks were eaten by a hawk before they ever got out of the nest.

Laurene Ratcliffe once spotted a small beak, a male, on Española, that was keeping two houses a few meters apart. In one cactus he was nesting with a female of his own species, and in the other he was feeding the nestlings of a warbler finch.

("This is a big controversy in Scandinavia right now," says Trevor Price. "A male has two nests. *Does she know?* That's all they talk about in Scandinavia. And it looks like she does know, but that's the best she can do.")

"Misimprinted birds are interesting because they broadcast not one identity but two," writes Peter Grant: "a false one at long distance and a true one at short distance." This is one way the finches can get their lines crossed and make hybrids. The misimprinted males tend to get into fights with males that sing their song—that is, with the wrong species—and they tend to breed with the wrong species too, when they mate at all.

EVEN THOUGH HIS THINKING about evolution was formed by what he had seen in the Galápagos, Darwin never quite understood how much species are like islands. The momentum of his long argument sometimes made him sweep past and ignore the distinctiveness of species, and talk as if all the categories of life run smoothly together. "I look at the term species as one arbitrarily given for the sake of convenience to a set of individuals closely resembling each other," Darwin writes in the *Origin*, as if species were just as arbitrary a convention as varieties, which are merely "less distinct and more fluctuating forms."

It is true that the thirteen species of Galápagos finches are things of the passing moment, like the bodies of the individual birds that make up each species, like the islands that are their home, and like the very planet that is the home of all these islands. But one of the most significant facts about bodies, islands, and planets is that while they last, they are real, distinct, and separate. True species are as real as bodies, islands, or planets, though what holds them apart is not as uniform or as obvious to the eye. Even now, even with hybrids flourishing triumphantly on Daphne Major, most of the finches on the island seldom interbreed.

Clearly song is part of the finches' secret. But song is not everything. If it were, the cactus finch that sings a *fortis* song would be able to mate with a *fortis* female, and the broadcasts of the loudspeakers that Laurene and Peter arranged on Daphne and other islands would have had an effect on both male and female finches. The males often attacked the speakers, but the females acted completely uninterested. In almost five hundred trials, only three females ever came over to the speakers.

To find out what was missing for the females, Peter and Laurene set up two speakers 10 meters apart on the lava and placed a stuffed male decoy on top of each one. Now the females were more interested. They hopped over to investigate each speaker. They came closer to a male decoy of their own species than a male of another, and spent more time near it. They also spent more time with a male from their island than a male from another.

Obviously the females are looking for something more than song. So are males, since when females approach a singing male, the females do not sing, yet the male decides somehow whether the female is one of his own species, a bird of his feather, worth the trouble and expense of courting.

The word species comes from the Latin verb *specere*, to see. Linnaeus used the word to mean groups of animals and plants that look distinctly different to the eye. After screening each other with songs, Darwin's finches must do their taxonomy as Linnaeus did. They must tell one another apart by eye.

To find out what the birds are looking for, Laurene Ratcliffe and Peter Grant performed a slightly ghoulish series of experiments. They collected dead finches from the lava, and also a few stuffed specimens from the museum at the Charles Darwin Research Station. They spruced up each corpse a bit and rejuvenated it with a spine of wire, so that they could bend the body into various poses. They also painted the beaks of the females dark brown and the beaks of the males black: mating colors.

They went to the heart of a male finch's territory, right near his nest. There they propped a female *scandens* decoy on one cactus pad and a female *fortis* on another cactus pad. They tried perching the female decoy in various poses, sometimes in sitting position and sometimes in the much more provocative pose in which the beak is angled upward toward the sky, the wings extended a bit, and the tail spread, an invitation to copulation. In the body language of the finches this is a powerful invitation. In fact when Ratcliffe and Grant first bent a model into this shape, male finches began courting it and trying to mount it while the stuffed bird was still in their hands. On Pinta, where the nests were quite high, Laurene would lash the decoys to a bamboo pole and then prop the pole up at the nest. When she carried the pole from territory to territory, males would swoop onto the decoys and try to mate with them. Laurene learned to cover the decoys with a bandanna until she had them ready.

Each experiment began when they took away the bandannas, unveiled the two rival female decoys, and backed away. Within minutes a male would come near and inspect the models. Usually, Ratcliffe and

Grant write, the male faced the decoy, "wing-quivering and whistling, and occasionally swaying from side to side." Then, within five or ten seconds, the male mounted and copulated with the decoy. Afterward, while still mounted, he sometimes touched his beak gently to the female's head or pecked it gently.

The finch watchers tested more than one species, and they repeated the tests on more than one island. Male finches most often chose the decoy of their own species. They did also show some sexual

Finches courting. He shakes his wings; she raises her beak into the air, a sign of interest.　　　　　　　　　*Drawing by Thalia Grant*

interest in females of alien species, although Grant and Ratcliffe noticed that very often in those cases the males' displays "suddenly collapsed in mid-performance."

Finally, in the most interesting and grotesque refinement of these experiments, Ratcliffe and Grant tried switching the heads and bodies of *fortis* and *scandens* on the decoys. They found that they could produce a smooth gradation of responses. A decoy with the head of a *scandens* and the body of a *scandens* got the most reaction from a *scandens*. A decoy with the head of a *scandens* and the body of a *fortis* got a smaller response. A decoy with the head and body of a *fortis* got the least response. Of course the males may have been put off by these weird-looking decoys. But if they were, there was no sign of it: their responses to the intermediate females tended to be more or less intermediate. They were not just going through the motions either: they left drops of semen on the decoys' tail feathers.

Many other species of birds have been shown to be able to tell one another apart on sight. But most of them seem to go by bold differences in markings and plumage: a red beak, a black head, a scarlet breast, Day-Glo yellow eye-rings. Not many birds rely on differences in size and shape as subtle as the cues that Darwin's finches are using.

The Grants would like to carry out more elaborate experiments, but the regulations of the National Park Service of Ecuador, and the Grants' own feeling for the birds, forbid them. Based on what they have seen in the wild, the species are kept apart by two barriers, one at a distance and one close up, like a moat and a wall. The first hurdle is song. The second is the close-up visual inspection. And here the cadaverous experiments show that something about the heads of the birds is crucial.

Now, there is nothing distinctive in the sizes, shapes, or feathers of the finches' heads. There is only one character above the neck that varies distinctively from species to species. The conclusion is inescapable: the feature that makes the finches most interesting to us is also the feature that makes them most interesting to each other. When they are courting, head to head, making decisions that are fateful for the evolution of their lines, Darwin's finches are studying the same thing as the finch watchers. They are looking at each other's beaks.

UNLIKE COASTLINES OF COLD lava, instincts are malleable. Natural selection can shape sexual preferences. So mating patterns can shift, change, evolve, like the beak of the finch.

Experiments in artificial selection have created spectacular and sometimes rather monstrous transmutations of sexual instincts. In one lab, investigators raised fruit flies of the species *Drosophila subobscura* in total darkness. After fourteen generations, the males no longer gave the typical courtship dance. Instead they tapped around in the dark until they found a female and then tried to force copulation; the females did not turn them away. "It takes just fourteen generations to turn a *Drosophila subobscura* from a courtly dancer to a blind, tapping rapist," notes the student of evolution James Shreeve. "How perishable are our essences! No doubt Aristotle would find all this fiddling around with the edges between things to be very dark stuff indeed."

The origin of species is the origin of the invisible, isolating walls that arise between two populations and make them living islands. Individuals vary in the characters that affect their attractiveness at mating time, just as they vary in characters that affect their capabilities to feed or fly. Variations in these characters can lead to sexual isolation, as with the poor bird that sang the wrong song. Sometimes variation can grow broader, producing what have been called "sex races." Gypsy moths in Japan and Korea are divided into races like these. Cross a male from one race and a female from another, and you may get females with the bodies of males, or whole cohorts of moths that look androgynous and are completely sterile. The sex races shift from one to the next, and to the next, as you travel south and west from Hokkaido to Honshu to Kyushu. Here we seem to see invisible barriers forming that can carve one species into many.

The drosophilist Ken Kaneshiro has made a special study of the courtship patterns ("often bizarre") of Hawaiian *Drosophila*. Kaneshiro calls these flies "the birds of paradise of the insect world." Like Darwin's finches they make natural subjects for the study of evolution in action. There are more than seven hundred species on the islands, and more new species are being collected all the time. The Hawaiian Drosophila Project is one of the most comprehensive efforts in the history of zoological classification.

Kaneshiro argues that sexual selection is "a powerful force in initiating the speciation process." Often the chromosomes of two species of Hawaiian fruit flies look almost identical, yet their genitalia are strikingly different. Kaneshiro and his colleagues are now looking more closely at the genes at the molecular level to try to find the genetic changes that make the difference. Sexual selection seems to drive these divergences, Kaneshiro argues, because "all the sometimes bizarre secondary sexual characters found in the males are used in some way during their complex sexual displays."

The flies' extreme genetic similarity may suggest that these species are very young, that they evolved very recently. If so, the pressures of sexual selection may sometimes act much faster than the pressures of environmental change—at least, to start the ball rolling, to make a branch start to split and fork.

Evolutionists have argued about sexual divergence for years, as they have about competitive divergence, and this argument runs almost exactly parallel with the former. Does the divergence in sexual taste occur while the species or varieties are apart, separated by physical barriers like cliffs, valleys, and arms of the sea? Or do sexual tastes diverge when the varieties are neighbors—are they driven apart by selection?

Now, in the Galápagos, the finch watchers know that the pressures of sexual selection can sometimes compare with the intensity of natural selection. "And to the extent you've got isolation, it's based on beak shape," says Dolph Schluter. So both divergences seem to be linked in these finches. Both kinds of Darwinian selection, natural and sexual, have gotten fixed (at least partly) on the birds' beaks. And both kinds of selection seem to be changing now, judging by the rise of the hybrids.

We think of sexual preferences as constant and unchanging through the generations. But now we really have to ask on what grounds we assume they are constant. We know there is variability—in some cases extreme variability—from one individual to the next, both in sexual tastes and in sexual characters. We know that these features are heritable, passed on from one generation to the next. We know that they are subject to powerful selection pressures. And we know from experiment that selection pressures can produce rapid changes from generation to generation.

So on what basis do we assume these things are more or less stable, like the rocks in the stream, and not merely waves and ripples in the stream? Arrangements between males and females—arrangements and behaviors we think of as primary, as given, fixed, almost as immutable as naturalists before Darwin considered species—these are not permanent at all. Behavior is the product of forces, contending forces that are still contending today, struggling within each generation. The borders between species are continually tested and redefined by the outcome of each member of each generation's luck in love—an amazing thought.

Darwin never understood how the intertwining of the two kinds of selection he put on the map, natural and sexual selection, could lead to the creation of these invisible coasts. He wrote relatively little about the subject in the *Origin* (despite the title). Nor did he treat it at any length in *Natural Selection*. "The twenty-five pages on species and speciation in his unfinished big book manuscript," the evolutionist Ernst Mayr wrote recently, "contain so many contradictions that they are almost embarrassing to read."

Now the picture is becoming clearer, with the help of those who survey both coasts at once, the visible and the invisible, and it is an extraordinary picture. All these forces, a wide field of conflicting forces, play like winds across the adaptive ranges that hold each species apart, or like storm winds at sea. The borders between species are as fluid and adaptable, as sensitive to changes in pressure, as the heaving waves in a high sea. And winds can split the waves, as if splitting the mountains or sending a new mountain or new archipelago up above the rest.

What drives the first widening wedge? It is (to switch metaphors) a little like the splitting of an amoeba: one population goes one way, and one goes the other. You have one vessel, one gene pool, and you end up with two. And the beginning of the split can be a very small thing. A detail can make the difference. Even a detail that has no adaptive significance can make all the difference in the world. In other words, the origin of species can lie in the kinds of small, subjective decisions and revisions that in our species come under the heading of romance.

We are exquisitely sensitive to one another's features, and they can be more fateful than we dream. "Cleopatra's nose, had it been shorter, the whole face of the world would have been changed," says Blaise Pascal in his *Pensées*. A little less from bridge to tip (or a little more) and Julius Caesar and Marc Antony might not have fallen in love with her. If Cleopatra's nose had been shaped a little less like the Grecian ideal, and a little more like Cleopatra's Needle, there would have been no Alexandrine war, no sea-fight at Actium. The whole arc of the Roman Empire would have been reshaped by Cleopatra's beak.

When Darwin met Captain FitzRoy for his job interview, the captain took an instant dislike to Darwin's nose. The captain was an amateur phrenologist and physiognomist, and prided himself on his ability to judge the character of his men by their skull bumps. FitzRoy felt sure that he was looking at the nose of a lazy man. He almost sent Darwin home. We might have lost the *Origin* and *The Descent of Man*. The whole face of human thought was almost changed by Darwin's beak.

ON THE ISLAND of Genovesa, sheltered by Darwin Bay, there is a lagoon that is one of the loveliest and even sexiest spots in the Galápagos. It is a pale blue pool of clear water and white coral sand, with no leeches and no sharks. Masked boobies gather in the bushes that fringe the lagoon, and perch on the lava bluffs above it, among the prickly-pear trees. Great frigatebirds with deflated red balloons (gular sacs) dangling from their chests fly overhead. In courting season these males stand around together among the salt bushes. Each time a female flies by, they shake their wings, puff out their shockingly red balloons, throw back their heads, and howl.

The Grant family camped by this lagoon for years. Nicola and Thalia may be the only human beings ever to have been more or less raised in this strange paradise, beneath the phantasmagoria of booby cries and finch songs, and the big wing-rustles of circling frigatebirds. Here they read *The Swiss Family Robinson*—and hated it. They thought it was ridiculous that everything the family needed simply washed ashore. They liked *Robinson Crusoe* better, because that was more like life in the clamorous solitudes of Daphne and Genovesa. The girls built driftwood rafts and did their homework in the middle of the lagoon (home schooling on Genovesa). Laurene Ratcliffe remembers Nicky one year playing the violin to the lava gulls. And it was here at the lagoon that Thalia learned to draw.

The Grants first began camping here in 1978, the year after the great drought. They pitched a tent on the cliff above the lagoon and began to measure variation in the beaks of the large cactus finches. As part of their standard measurements they looked at the two groups of singers on the island, the A singers and the B singers. The As were singing variations on a theme that sounded like *chuh-chuh-chuh*. . . . The Bs were singing variations on a shorter theme that sounded like *chrrrrr.*

That first year by the lagoon the Grants were surprised to find that the As and Bs differed slightly in their beaks. The As' beaks were on average narrower, shallower, and longer than the Bs'. The difference in length was only about a millimeter, but of course a millimeter can make a world of difference. In fact that year in the dry season they saw the As, with their longer beaks, drilling holes in cactus fruits and pulling out the seeds. They never saw any of the Bs doing that; instead the Bs were concentrating on fallen cactus pads, ripping and tearing them open with their beaks, eating the pulp, and picking out the grubs. The Grants never saw any of the As doing *that*. So when it came to feeding, the single millimeter's difference was forcing the two groups of singers somewhat apart.

No one had been posted to watch the finches on this island the year before, during the drought. But the drought had struck here, as it had on Daphne Major, and although the Grants could not be sure, the stress of the drought might have caused this disruption in the large cactus finches, much as it had caused a split in *fortis*, the medium ground finches, on Daphne. In that split, *fortis* with the very biggest and the very smallest beaks were favored. The biggest were favored because they could handle the big seeds that are the province of *magnirostris*; a few of the very smallest female *fortis* were favored because they were able to poach the smallest seeds, the niche of *fuliginosa*.

So the drought might have caused a split here too, although no one had been here to watch it happen. The Grants found another surprise. Among the mated males there was not a single pair of neighbors singing the same song. Yet among the bachelors, the unmated males, the territories were a mix, sometimes A next to A, sometimes A next to B. This suggested to the Grants that the females were choosing to settle down only with males whose neighbors sang the other song type.

When the first rains come to the Galápagos, large females are ready to mate a few days earlier than small females. These large females, who get first choice of the males, tend to choose males whose neighbors—whose nearest rivals—are *not* singing their song. Now there were more Bs than As. So the As stood out and were most attractive to the big females. In this way, the year before, during the drought, more big females might have mated with the A singers, and their offspring might have had larger beaks than the Bs'. Then in the intense drought the differences in beak length could have been widened further by disruptive selection. Those with the longest beaks could drill the fruits best, and those with the shortest could rip the pads best. Selection would favor the longest and the shortest beaks and widen the difference between them.

The Grants will never know for sure if that is what really happened. They still kick themselves. If only they had been at the lagoon the year before! An evolutionist has written, "It is a fundamental difficulty of an historical science like the study of evolution that one can never establish the cause of a past event." In this case the past event was less than a year before.

Whatever had happened during the drought year, the Grants could see that the population had split slightly apart as if by the thin edge of the wedge. The separation in breeding also showed up in a more direct way, in the nest. When Darwin's finches first hatch, their beaks are always either pink or yellow. It is much the same with barnyard chickens. The beaks of a brood of hatchlings may be all pink, all yellow, or some of one and some of the other. The color lasts for their first two months of life.

As the Grants went around banding nestlings that year, they noticed that the nests of the song A males had more than twice as many yellow-beaked chicks as the nests of the song Bs. If the two groups had been mating and mixing at random, that would not have happened; they would have fathered roughly equal frequencies of yellows and pinks. So it seemed the gene pools of the As and the Bs were partly divided too.

Thus the Grants were looking at two groups that showed signs of developing the kinds of fixed differences in beaks, songs, and survival skills that mark the separate species of Darwin's finches. The difference in the beak measurements of the two groups, the As and the Bs, was about 6 percent. The difference in separate species of Darwin's ground finches averages about 15 percent.

"Subdivision is incipient speciation," the Grants write, and then add, with their usual caution, "or at least it provides the potential for speciation." At the time they thought that might be what they were about to see.

Unfortunately, the dry season of 1978 was very dry on Genovesa. Of the 120 nestlings the Grants banded by the lagoon, only six were still alive a year later. So the Grants could see that they would need to keep watch a long time on Genovesa, as on Daphne, to understand what might be going on.

It would not have taken much for the split between the As and the Bs to widen. The daughters of these two lines would have to keep on choosing males that sang the same song as their fathers. Then the division would have kept on growing.

But the next year the females around the lagoon chose males that sang the opposite song as their fathers as often as they chose males who sang the same song as their fathers. They chose males with the long form of the beak as often as males with the short form. (Maybe a difference of only a millimeter is not enough to make or break a couple.) Because they interbred this way, the beaks of their sons and daughters were in between the beak lengths of the two lines, and the tantalizing differences in beak length disappeared and fell apart. Their offspring no longer ate in two different ways, and they were as likely to have yellow or pink beaks as anybody else.

By 1979 the checkerboarding of the territories of the As and Bs was gone, although it may have appeared again briefly once or twice, just for a week at a time, in later years. (The Grants feel sure it did; but to a disinterested observer their evidence seems softer than usual here. The boundaries of birds' territories are harder to measure than the shapes of their beaks, and of course the Grants were looking very hard to see if that checkerboard would come back.)

In any case, by 1981 the link between the shape of the beak and the songs of the As and the Bs was completely gone. A millimeter still made the difference between the fruit borers and the pad rippers, but the birds no longer divided into As and Bs.

In 1985 another drought came to Genovesa, and this time the Grants were there to watch. Not a single drop of rain fell that year. But this time, the drought did not partition the birds into two groups, as the Grants believe it may have done in 1977. Of course the island was a different place in this drought than it was in the one before, because of the crazy Niño of 1982–83. Most of the cactus trees had been overrun and toppled by vines. There were hardly any fruits and seeds to be had anywhere in anyone's territory. So the pressures of the drought favored short, deep beaks for ripping up dried cactus pads, but there was no separate niche for skinny, long beaks. The drought of 1985 pushed the whole population in the direction of the only niche that was open to them. The survivors had significantly deeper beaks.

In a sense the future evolution of this population of cactus finches is cramped by its neighbors. These cactus finches are not alone by the lagoon. There are sharp beaks to pick up the small seeds, and there are big beaks to pick up the really big seeds. So the cactus finches are hemmed in. There is no big new empty niche waiting for a group of them, like adventuresome colonists, to split off and explore; at least, there is no niche the Grants can see. These finches have been on

Genovesa for a long, long time. There is no new world out there by the lagoon that is just sitting there, ripe for the taking.

Thus the Grants suspect that the finches here are perpetually being forced slightly apart and drifting back together again. A drought favors groups of one beak length or another. It splits the population and forces it onto two slightly separate adaptive peaks. But because the two peaks are so close together, and there is no room for them to widen farther apart, random mating brings the birds back together again.

These two forces of fission and fusion fight forever among the birds. The force of fission works toward the creation of a whole new line, a lineage that could shoot off into a new species. The force of fusion brings them back together. What the Grants saw was the population cinching at the waist like an amoeba in the act of dividing, and the division was just a millimeter in the birds' beaks.

Chapter 12

Cosmic Partings

I was of three minds
Like a tree
In which there are three blackbirds.

— WALLACE STEVENS,
"Thirteen Ways of Looking
at a Blackbird"

In the *Origin*, Darwin gives a sketch of a few growing twigs on the tree of life. It is the only illustration in the book, and it is highly schematic and abstract. A dozen lines of life, labeled A through L, rise upward through a series of horizontal lines numbered I through XIV.

These horizontal lines represent "long intervals of time," Darwin says, and as he talks about the diagram he keeps stepping farther and farther away from it, contemplating wider and wider expanses of time.

"The intervals between the horizontal lines in the diagram, may represent each a thousand generations," Darwin writes, "but it would have been better if each had represented ten thousand generations. . . ."

"In the diagram, each horizontal line has hitherto been supposed to represent a thousand generations, but each may represent a million or hundred million generations, and likewise a section of the successive strata of the earth's crust. . . ."

What Darwin does *not* do, either in this "Diagram of the Divergence of Taxa" or anywhere else, is to step closer and contemplate the moment of divergence itself, the point where one line splits in two, where two roads diverge, the cosmic parting.

He could not look at the point of origin in any particular case, because he did not have one to offer; he did not try to contemplate the mechanics of divergence even in the abstract, because he did not know enough.

Today, as more investigators take up the study of evolution in action, they watch for these cosmic junctures, the point where lines diverge, the parting of ways. What has already emerged from these studies is how fast divergence can happen, how much we can actually see in real time. We are not talking about the lapse of ages.

Some years ago, for instance, the evolutionists Theodosius Dobzhansky and Olga Pavlovsky published a now-famous article in the *Proceedings of the National Academy of Sciences* entitled "Spontaneous Origin of an Incipient Species in the *Drosophila paulistorum* Complex."

"It has been questioned, by His Holiness Pius XII among others, whether biology has really succeeded in making a species from another species," the evolutionists began. The process of speciation "is generally too gradual and slow to be observed directly. An exceptional situation . . . is reported in the present article."

The *Drosophila paulistorum* complex is what is known as a superspecies. It is a cluster of populations that are partly but not completely isolated from one another by their build or their behavior—like Darwin's finches, but more promiscuous, interbreeding more freely. The *paulistorum* complex consists of six populations of fruit flies that are, in the words of Dobzhansky and Pavlovsky, "too distinct to be considered races of the same species but not distinct enough to be regarded [as] full species." Their ranges lie across much of Central and South America, from Guatemala to the Andes.

A superspecies is just the kind of messy, borderline case that fascinated Darwin. Here is a situation in which, as Dobzhansky once wrote, "the process of species splitting has, on our time level, reached the critical stage of transition from race to species. *D. paulistorum* is one species; it is also a cluster of species *in statu nascendi;* it bespeaks the correctness of Darwin's opinion that 'each species first existed as a variety.' "

Within the *paulistorum* cluster, Dobzhansky and Pavlovsky explained, female fruit flies will mate freely with males of their own incipient species, but seldom with males of another. That is, an Amazonian female will mate with an Amazonian male but not often with a Guianan, or an Orinocan, or any of the other incipient species in the complex. When flies in the *paulistorum* cluster interbreed, which they do more often in the laboratory than in the wild, they tend to have fertile daughters and totally sterile sons.

All these flies look alike. To diagnose an unknown strain, Dobzhansky and Pavlovsky had to cross it with what they called "testers," a series of flies representing the whole array of incipient species in the *paulistorum* complex. The unknown strain would yield partly sterile progeny, or none at all, with most of the testers. It would give completely fertile hybrids with only one set of testers: those of its own kind.

"On March 19, 1958, a sample of *Drosophila* was collected at Chichimene, south of Villavicencio, in the Llanos of Colombia," Dobzhansky and Pavlovsky reported in the *Proceedings*. From these flies the evolutionists reared a strain called Llanos-A. (*Llanos* are the wide grassy plains that still cover much of South America, as in the days when Darwin rode over them on his long inland wanderings from the *Beagle*.) Because the Llanos-A strain yielded fertile hybrids with most Orinocan strains, Dobzhansky and Pavlovsky classified Llanos-A as a member of the Orinocan incipient species. The evolutionists kept Llanos-A alive in their laboratory at Rockefeller University, in New York City, as well as strains from all of the other flies in the *paulistorum* complex.

Five years later, as part of an unrelated study, the investigators began testing all these strains again. By now, generations of fruit flies had come and gone within their separate population cages. But all of the investigators' strains yielded the same results as before—all of the strains, that is, but one. "The Llanos-A strain now behaved in a totally unexpected manner; namely, it failed to give fertile hybrids with any strain other than itself," Dobzhansky and Pavlovsky reported. With Llanos-A, no matter which testers they tried, Dobzhansky and Pavlovsky could not make a single fertile cross. Llanos-A was no longer compatible even with Orinocan strains, the very same strains with which it had produced fully fertile hybrids only a few years before.

"*Conclusions*—Llanos-A is a new race or an incipient species having arisen in the laboratory at some time between 1958 and 1963. It has diverged from its progenitor, Orinocan. . . ."

No one was watching in the interval, so no one saw it happen. Why did that strain diverge? Dobzhansky and Pavlovsky suspected that the flies had become infected with a bacterium. The infection might have rendered the flies infertile with any strain but their own. This probably is just what happened: infections of that kind have recently been detected and confirmed by drosophilists in several laboratories.

The infection sweeps through the population cage fast, like cholera or flu, and it can carve invisible borders around a population, wall it off from all the rest of its kind, like a new and invisible coastline or an invisible cage, virtually overnight. Most of these infections can be cured with antibiotics, some not. In principle the same thing could happen to our own kind, or make all but a small, lonely pocket of the human population sterile, as in Kurt Vonnegut's wonderful novel *Galapagos*. If the divergence of Llanos-A was driven by a bacterial infection, then it may have happened literally overnight.

AS WITH THE ORIGIN OF SPECIES, so with the origin of adaptations. A recent experiment suggests that this process need not be as gradual as Darwin imagined. It also shows how tightly the origin of species and adaptations are linked together.

The origin of adaptations is, like the origin of species, one of the deepest problems in Darwinism. How do novel adaptations arise from small and gradual beginnings? The ancestral line of Darwin's finches was all of a kind. One set of finches blew into the islands, and today we have a finch that perches in trees, making and wielding toothpicks against grubs, and another finch that perches on the backs of boobies with its long, sharp beak dipped in their blood.

How did blind creation make so many new kinds of tools? How do evolutionary inventions and innovations like these get started, if their raw material is merely random individual variations? As one of Darwin's early critics writes, it is hard "to see how such indefinite oscillations of infinitesimal beginnings can ever build up a sufficiently appreciable resemblance to a leaf, bamboo, or other object, for Natural Selection to seize upon and perpetuate."

No one has ever put this problem more forcefully than Darwin himself, in the *Origin*. "To suppose that the eye with all its inimitable contrivances for adjusting the focus to different distances, for admitting different amounts of light, and for the correction of spherical and chromatic aberration, could have been formed by natural selection, seems, I freely confess, absurd in the highest degree," he writes. But if we look at the whole tree of life, Darwin says, we can find innumerable gradations from extremely simple eyes consisting of hardly more than a nerveless cluster of pigment cells, which are rudimentary light sensors, to the marvels of the human eye, which are more impressive pieces of work than the human telescope.

Darwin argues, essentially, that all the sophistications we see in the eagle's or the human's eye could have arisen gradually, by stages, across geological spans of time, each stage conferring somewhat clearer vision than the one before. "We must suppose each new state of the instrument to be multiplied by the million; each to be preserved until a better one is produced, and then the old ones to be all destroyed. . . . Let this process go on for millions of years; and during each year on millions of individuals of many kinds; and may we not believe that a living optical instrument might thus be formed as superior to one of glass, as the works of the Creator are to those of man?"

Darwin was emphatic that all complex adaptations arise by the gradual agency of natural selection. He even makes the point a sort of test case: "If it could be demonstrated that any complex organ existed, which could not possibly have been formed by numerous, successive, slight modifications, my theory would absolutely break down."

Richard Dawkins defends Darwin's position vigorously in his book *The Blind Watchmaker*, which is framed as a reply to the Reverend Paley's parable of the watch on the heath. Dawkins argues that selection on even the slightest nubbins and rude beginnings can build up instruments as complicated as watches, telescopes, or human eyes. As long as each stage in the evolution of a complex adaptation is adaptive for its own sake, it is likely to be preserved within each generation and be embellished by the next, by Darwin's process of natural selection. The process does not look ahead. The watchmaker is blind. Yet blind selection can build an eye. Suppose, Dawkins writes, a few nerve endings in a simple organism gave it a rudimentary sense of light and dark. Even rudimentary vision is better than no vision at all. The variant individual that carried that adaptation would be likely to be preserved; and so would those that inherited those first dim eyes. "Vision that is 5 percent as good as yours or mine is very much worth having in comparison with no vision at all," writes Dawkins. "So is 1 percent vision better than total blindness. And 6 percent is better than 5, 7 percent better than 6, and so on up the gradual, continuous series."

The evolutionist Stephen Jay Gould notes that a big change in an organism can sometimes arise from a small change in its genes. So adaptations may sometimes evolve by big steps as well as small steps. Suppose a random change in the genes is sizeable, but not so big that it prevents the animal from breeding with others of its kind. "Suppose also," Gould writes, "that this large change does not produce a perfected form all at once, but rather serves as a 'key' adaptation to shift its possessor toward a new mode of life." If it began to make a living in a new way, it would find itself subject to a whole new constellation of selection pressures, just as if it had been picked up by a storm and dumped down on a desert island.

In his recent book *Darwin on Trial*, the lawyer Phillip E. Johnson speaks sarcastically of "all this supposing." "Gould supposes what he has to suppose, and Dawkins finds it easy to believe what he wants to believe, but supposing and believing are not enough to make a scientific explanation," Johnson writes, adding, "The prevailing assumption in evolutionary science seems to be that speculative possibilities, without experimental confirmation, are all that is really necessary."

There is now a simple experimental confirmation of this point. The experiment was published the same year as Johnson's book, 1991. It was carried out by two evolutionists at the University of British Columbia, in Vancouver, working in a borrowed corner of the laboratory of Dolph Schluter.

There is a genus of finches with peculiar mandibles that cross over at the tips. There are about twenty-five species and subspecies of these birds in North America, Europe, and Asia. They are called crossbills, or crossbeaks. According to legend, they twisted their beaks when they were trying to wrest the nails from Christ's cross. The red on the breast of the male is the blood of Christ.

Darwin was intrigued by the beak of the crossbill. In *Natural Selection* he notes how variable they are "in length, curvature & the degree of elongation of lower mandible." He notes further that some small curvature of this kind had been reported in many species of birds, as a deformity, and that in good times some of these deformed birds survive.

The peculiar mandibles of these finches are adaptive. As Lack notes in *Darwin's Finches*, there is a small-beaked crossbill that feeds chiefly on the soft cones of the larch, a medium-sized crossbill that feeds on the harder cones of the spruce, and a heavy-beaked parrot crossbill that feeds on the still harder cones of the pine. The twisted beak allows the bird to pry open closed cones. The connection between the beak and the food is so obvious that it was accepted by evolutionists even at the time Lack wrote, when his colleagues believed that most differences between sibling species have no adaptive significance at all.

The evolutionists Craig Benkman and Anna Lindholm performed their experiment on seven captive red crossbills. The red crossbill lives

Three crossbills and their special interests. Top, a parrot
crossbill with a pinecone. Middle, a common crossbill with a
spruce cone. Bottom, a two-barred crossbill with a larch cone.
From Ian Newton, *Finches*. Courtesy of HarperCollins
Publishers Limited.
Library, the Academy of Natural Sciences of Philadelphia

in coastal forests from Alaska to California, and its beak is specialized
for the cones of western hemlock. What Benkman and Lindholm did
was to uncross the beaks of these birds by trimming the crossed part
of the mandibles with an ordinary nail clippers. This did not hurt the
birds, because they have no nerve endings in their beaks: it was as
painless an operation as trimming claws or fingernails.

The birds with uncrossed bills turned out to be just as good as ever
at extracting seeds from dry, open pinecones. But they could no
longer handle closed cones. Of course they still tried, the way a
declawed cat will still try to climb a tree, but with their straightened
beaks they could not get anywhere. Day by day, as the twist in their
beaks grew back, the birds did better and better with more and more
recalcitrant cones. After a month, their beaks were completely re-
grown, and they were back in business.

What is striking about this little experiment is that Benkman and Lindholm could measure the value of an adaptation from its very beginnings to its final form. If crossed mandibles were useful to these birds only when fully formed, then it really would be a puzzle how they could have arisen by natural selection. The cross would have to appear all at once, as what the geneticist Richard Goldschmidt called "a hopeful monster." It would be the kind of problem before which Darwin felt his theory would "absolutely break down." But the finches began to get better at opening pinecones when the cross in their beaks was still too small to be visible to the eye. Even a slight crossing of the mandibles confers a small, incremental benefit, making more and more tightly closed cones accessible. So it is easy to see how the crossbill's crossed bill could have arisen gradually, by selection, over generations, each generation doing a little better with closed cones than the generation before. The press of competition in the woods would have made the novelty of a crossed beak more and more desirable, because it would allow its possessor to eat food no one else could eat; the same competitive pressure would favor each new twist. New worlds kept opening around the birds: pinecones, spruce cones, hemlock cones, fir cones. Today, however, there is no profit to a sparrow or bunting in a deformed, twisted bill, because the crossbill niche is taken.

One new adaptation opened a new way of life, and led to a host of other adaptations, including refined instincts for cone hunting and strong muscles to operate the peculiar beak. Today these crossbills are so skilled at opening cones and so committed to their twisted way of life that they eat almost nothing else. In good years their specialty sets them apart and gives them a diet that is closed off from all the other birds in the woods. But when a crop of pinecones fails, they often starve to death.

A small, simple variation in a beak has led to an adaptive radiation that is, counting by the number of living species and the number of individual birds, much larger than the divergence of Darwin's finches. The finches in the forests did not have to discover a new archipelago. Two roads diverged, they took the road less traveled by, and that has made all the difference.

IT WOULD BE EVEN MORE SATISFYING to watch the process of divergence in action, in real time, from beginning to end. "Wouldn't it be lovely," Dolph Schluter asks, "if you could actually introduce *fuliginosa* to Daphne—and track the changes for the next one hundred years? That is inconceivable, of course. Or, alternatively, if you could remove *fuliginosa* from islands that have both *fuliginosa* and *difficilis*, and track the evolution of beak size for the next one hundred years and more. That would be a wonderful thing!"

No one will ever do that, because no one can wait one hundred years, and also because the finch watchers and the directors of the National Park Service of Ecuador have too much respect for the birds. As Dolph puts it, "These populations are so unique that we have no desire at all to mess with them."

So Dolph has begun a new study, not far from Mandarte Island, in the Strait of Georgia. The Strait of Georgia is full of three-spined sticklebacks. These sticklebacks live along most of the ocean shorelines of the Northern Hemisphere. They swim into the mouths of rivers and streams, thousands upon thousands of inlets and arms of the sea, wherever the tide of salt water meets the current of fresh water, to breed.

Sometimes these small salt-water fish ended up invading fresh water and coming to stay. In southwestern British Columbia, as the ice melted from the land at the end of the last ice age, sticklebacks swam into some of the new glacial lakes at the edge of the sea, and got stranded there. "All of the lakes are younger than 13,000 years. That sounds like a long time, but it's—lightning," Dolph says, laughing. The lakes were cut off from the sea about 12,500 years ago. For 12,500 years now the fish have been evolving in those lakes like finches marooned on islands.

Most of the lakes in that part of British Columbia hold just one species of stickleback apiece. But in the last few years, Schluter and John Donald McPhail (a longtime student of sticklebacks) have discovered that lakes on the islands of Texada, Lasqueti, and Vancouver each hold a pair of species.

These pairs are so new to science that they have not yet been named, but they fall into two general types. In each pair, there is a species that feeds along the bottom, and a species that feeds above it, up in the water column. Dolph and others call the two kinds of fish "benthics," from the Greek word *benthos*, the depths of the sea, and "limnetics," from *limnos*, marsh, which by modern usage refers to life in the open waters of lakes, away from the shore.

These are wilderness lakes, and the sticklebacks in them are almost as free and wild as the finches of the Galápagos. The finches' only regular enemies are owls and hawks; the sticklebacks' only enemies are cutthroat trout. So what Dolph has here is a whole new set of test tubes for the same experiment.

"I want to do it again!" he says.

Dolph has looked at the pairs of sticklebacks in Enos Lake, on Vancouver Island; in Hadley Lake, on Lasqueti Island; and in Paxton, Priest, and Emily lakes, on Texada Island. The largest of the lakes is 44 hectares, which is about the size of Daphne Major, and the smallest is 5 hectares, about the size of Darwin's old property at Down House, not counting the Sandwalk. For a few years now Dolph, Don McPhail, and their team have been sampling these lakes with minnow traps and seines, and measuring the sticklebacks. The team measures body length, body depth, and width of the gape, and the length and number of gill rakers.

The gill rakers are finger-like digits that filter the food the fish swallows. Their size is strongly related to the size of the fish and the size of the food it can eat. Coming from the Galápagos, Dolph says, he likes to think of the gill raker as the beak of the fish.

Like Darwin's finches, these fish are variable; and like the finches, two of their characters are extremely variable: their body size and the length of their gill rakers.

To get to know the fish and their diets, Dolph has had to learn to recognize on sight a whole new bestiary of animals and plants. The fish down at the bottom of the lake eat annelids, amphipods, gastropods, pelecypods, harpactacoid copepods, and chydorid cladocerans, among other tidbits. The fish that swim up and down the water column eat a whole different flora and fauna. There is almost no overlap in their diets.

Dolph and the other stickleback watchers have measured the length and width of every one of these foods. The bits of fish food are so small that Dolph has to look at them through a dissecting microscope to measure length and width.

With each and every pair of sticklebacks, the pairs are divergent, and every pair is divergent in the same way. One species in the pair always feeds on the bottom, the other always feeds in the water column, excluding the bottom. Moreover, in every lake, the benthics tend to be bigger and fatter, with bigger mouths and wider gapes but shorter and fewer gill rakers. The limnetics are smaller, thinner, with narrower gapes but longer and more numerous gill rakers. The benthics tend to eat the biggest prey, the limnetics the smallest prey.

Five lakes, five pairs of species, and one pattern, repeated again and again. Meanwhile the sticklebacks that have a lake to themselves are intermediate in size, shape, gape, and gill rakers, and they range freely between the bottom and the water column, dining where they please, enjoying the best of both worlds.

Even within the solitary species the pattern repeats. "In Cranby Lake, which is only a stone's throw from Paxton Lake, there is only one species," Dolph says. "But what's amazing is when you look at individuals, there it is again." The great majority of the fish that Dolph has netted in Cranby are specializing, taking at least 90 percent of their food either from the bottom or from the water column; their choice of specialty depends strongly on their build. The bigger ones with the shorter gill rakers are down at the bottom, and the smaller, slenderer fish with the longer, more numerous gill rakers are swimming up and around in the water column. So tiny variations in the dinnerware make a difference in the way the individual fish live their lives; the shape of the gill raker is as vital to them as the shape of the finch's beak. The difference in the length of the gill rakers, the decisive difference that consigns a stickleback to one or the other of these two worlds within the lake, is only a third of a millimeter.

"Variation is there already. It's ever-present," Dolph says. And as among Darwin's finches these variations are passed down through the generations. Almost half the variation in gill-raker length is in the genes. "That's very high," he says. "In other words, there's lots of genetic variability. The message being, these fish as we see them now are not stuck in a rut. If the environment changed they could respond. These populations are ready to respond should selection pressure arise."

So there is a simple trade-off here for a stickleback. If the fish specializes in the muck, it cannot compete in the open water; if it specializes in the open water, it is outclassed down in the muck. The fish is in much the same position as a finch in the Galápagos, where specializing in big seeds unfits you for the small ones, and specializing in the small seeds unfits you for the big ones.

To Dolph all this evidence powerfully suggests that the colonists in these lakes have altered the course of each other's evolution, just as the finches have altered each other's courses in the Galápagos. They are a new case of character displacement, as neat in their way as the Galápagos finches, and Dolph found them right in his own backyard.

The reason the pairs of species have divided their lakes the same way, over and over again, is apparently that natural selection creates the same intense pressures on the fish in every one of the lakes. In each lake, with almost no competitors around except the other sticklebacks, the quickest way to get away from the competition with the least sacrifice is to go up or down the water column. And that is what these pairs appear to have done, again and again, until they look almost as stratified as two layers of rock. But unlike layers of rock they are alive, and their impressive variability implies to Dolph that they are evolving now. The process of divergence is still going on today in those five lakes, for "natural selection continues to hold the species apart."

In principle, then, it should be possible to watch. If he is right, the adaptive landscape in the lakes of British Columbia has two big peaks with a valley between them, and where there are two species of sticklebacks in a lake, each takes one peak. Now Dolph wants to watch the sticklebacks evolve toward those peaks.

"If there are two species of sticklebacks in a lake, and they really are in mutual repulsion, then we should be able to take one out and see if the other moves toward the middle," he says. "These are rare species, and they are on Canadian threatened-species lists. So we don't want to screw around with these lakes. But we can construct our own lakes, or ponds, and introduce the species, and see what happens."

This is Dolph's answer to Connell's "Ghost of Competition Past." "Connell's idea is, if you see two species that are different, you are always able to fabricate an argument that selection formed those differences," Dolph says. "The idea in his mind is that what we see now is the consequence of selection that happened a long time ago.

"But you look at stickleback populations, how young they are and how large their differences are, it's quite clear that competition is no ghost. It affects populations today. And that's what we hope to demonstrate with these experiments.

"Connell's idea about competition past is related to the more general idea that evolution took place a long time ago, that it is history. Whereas, in fact, natural selection is out there," Dolph says, with that same mellow laughter in his voice. "One can actually see it happen within the time of a Ph.D. thesis. People never tried to look. They thought you would have to watch a population a thousand years. But that's changing now."

The ponds are 70 by 80 feet, about 10 feet deep at their deepest. Dolph has built thirteen of them on campus. "If you go to a lake and you see two species and they are quite different," he says, "the divergence argument implies that natural selection is maintaining those differences. Natural selection is acting all the time to maintain those differences. If you were to remove that pressure, they would relax toward the center. And this is what we intend to test."

While the Grants sort through their numbers in Princeton, Dolph is sorting through sticklebacks in Vancouver. He and the Grants still keep in touch, and he keeps them up-to-date about his evolutionary experiment. For now he is readying the ponds and breeding up batches of sticklebacks. "I can't wait," he says. "I've got fish coming out of my ears. The lab is full of them. They're eating me out of house and home. I can't wait to be rid of them."

In some of his ponds, Dolph will introduce limnetics; in other ponds, benthics. He predicts that in the solitude of its own pond, each species will evolve in the direction of the other. That is, the two species will converge over time, generation after generation, until their gill rakers are neither large nor small but medium-sized. Then all of the fish will be able to dine on the muddy bottom and also in the clearer waters above.

At the same time, Dolph is collecting sticklebacks from different lakes, some limnetic, some benthic, some in between. In his laboratory he is crossbreeding them to create a hybrid swarm—a line of sticklebacks that is far more variable in size and shape than any line in the wild. "Because they are more variable," he says, "they will allow more powerful and sensitive measurements of natural selection, as they are 'sampling' a broader reach of the adaptive landscape. I like to think of them as 'selection probes'—instruments designed to measure selection." He will stock a pond with these selection probes, add a line of benthic or limnetic competitors, and watch what happens.

"I'm sure we'll be able to measure selection pressures," Schluter says. "We've done it in the Galápagos, we can do it here. It might be longer before we are able to detect an actual evolutionary response. One year is one generation. I think we'll be able to do it in ten, anyway!

"Yeah, we've got our work cut out for us."

Chapter 13

Fusion or Fission?

The closer one looks at these performances of matter in
living organisms, the more impressive the show becomes.

— MAX DELBRÜCK,
A Physicist Looks at Biology

Rosemary sits on her backless chair with her chin in hand, staring at
the Macintosh. More martial columns of numbers parade down the
screen. "It's such a *lot*," she says flatly, without taking her eyes from the
screen. "And we've got to be so terribly careful. I mean, the amount
of cross-checking, and double-checking, and triple-checking, to make
absolutely sure there are no errors in the data . . . !"

Her office is quiet. Cactus finches probe cactus flowers in the photographs on the walls. Guppies hover in an aquarium on the window-
sill. They too are souvenirs of evolution in action, although no one is
watching them now. They come from the famous fish tanks in John
Endler's laboratory. Rosemary's daughter Nicola got them from some
of the grad students. "*She* wanted them, but then *I* of course ended up
with them," says Rosemary.

She and Peter have been climbing their mountain of numbers day
and night, here in Eno Hall, and on a Macintosh in their house on
Riverside Drive, a few minutes from the Princeton campus. They
work together on the same desert island in the Galápagos and in the
middle of civilization. She sometimes wonders if they could have done
it when they were first married. But now they know how to work
with each other, and they know how to work around each other.
They interlock.

This sabbatical they have carted much of their hybrid data across
the Atlantic on visits to Uppsala, and to Arnside, her parents' village
in the Lake District, then back again to Princeton. They have combed

through the hybrid data a dozen times and looked at the numbers from a dozen directions. Bit by bit, the heaps and mounds of data have grown tidier, shapelier, more manageable. On good days now it seems as if she has gotten up above the desert islet in a balloon, as if she and Peter have climbed a mountain above the mountain and they are looking down on all they have done and seen in the last twenty years.

THEY CAN SEE in their numbers and computer-generated charts that since the flood, the crazy Niño of 1982–83, the adaptive landscape of the islands has changed dramatically. *Tribulus* has gone down, down, down. It was already in trouble before the first rains of El Niño. Rosemary and Peter suspect a fungus: some kind of rust that ate at its roots. Then of course the floodwaters drowned it, and green vines overwhelmed it, and *Cacabus* plants sprang up as if out of nowhere—great, sticky, hairy-leaved mats of *Cacabus*—and smothered it. After that came the droughts.

Cactus on the island is down too. First the cactus trees took up too much water in the great flood, then too little in the droughts. The cactus trees toppled under tangles of vines and more *Cacabus*. By 1990 there were hardly any *Tribulus* or cactus seeds to be found anywhere on the island, even for finches flipping hours of pebbles with their beaks. The cactus may be just beginning to come back now.

A pattern is emerging: a shift that the Grants had not seen clearly until they did the analysis. There have been fewer big, hard seeds on Daphne since the flood. But there have been more small, soft seeds— mostly from all that *Cacabus*. Rosemary and Peter have put these changes into hard numbers, and the changes are significant. In fact, the changes are enormous. For Darwin's ground finches, life is seeds. If the pile of big seeds shoots down and the pile of little seeds shoots up, that is an upheaval, a catastrophe in the adaptive landscape, the fall and rise of alps. This is just what has happened on Daphne in the years since the flood. One adaptive peak has collapsed, and another peak has gone beetling skyward.

These changes have been especially hard on the cactus finches. Cactus is their only home in the adaptive landscape (and in the ordinary landscape). If cactus falls, they fall too. The Grants have plotted the numbers of cactus finches on the island in relation to the numbers of cactus trees, fruits, and seeds. They can see that during each of the droughts since the flood, the cactus finches' population on Daphne declined, as expected. By the start of this year, when Rosemary caught those two rogue finches on the north rim, there were only about one hundred cactus finches on Daphne, which is the lowest their numbers have dropped since the finch watch began.

In spite of all this selection pressure, the cactus finches have not changed in the last ten years. By all the Grants' measures, their beaks and bodies are the same now, on average, as they were before the flood. This too makes sense in terms of the adaptive landscape, because in evolutionary terms these birds have nowhere to go. "Flee as a bird to your mountain," sings the psalmist. Cactus is this bird's mountain. When this peak falls down, they have no peak nearby to which to flee. They are trapped on one falling alp.

Since the big Niño, selection pressure has been strong on *fortis* too. Among the *fortis* that saw the flood, not quite one in three were still alive by 1987. But the Grants' tables show that the *fortis* did not die at random. The survivors in 1987 were eating a much greater proportion of small seeds than large seeds. This was partly a change in behavior, since individual *fortis* are flexible in their choice of foods. However, they are flexible only up to a point. The Grants can see from their data base that it was the *fortis* individuals with significantly deeper, wider beaks, the birds that were committed by their anatomy to the pile of big seeds, the eroding peak, who were doing most of the dying. The *fortis* with significantly shallower and narrower beaks were doing most of the surviving. So the average beak of the *fortis* generation that was born after the flood—the baby-boom generation—was better adapted to the brave new landscape of the 1980s.

In other words, while the cactus finches have gone down with their peak, the *fortis* have evolved. They have rolled with the adaptive landscape. The width of the *fortis* beak in the new generation, a generation of finches that is hopping around on the lava of Daphne Major at this moment, is measurably narrower than the beaks of the generation before them—down from 8.86 millimeters at the time of the flood to 8.74 millimeters now.

That does not take *fortis* back where they were at the start of the Grants' watch, but nearly so. The birds shifted toward large size in the first years of the study, and now they have shifted most of the way back. It is as if the whole island has dodged back and forth beneath its load of perching birds, as lost Spanish sailors once believed the whole archipelago could do, which is why they spoke of *Las Encantadas*, the Enchanted Islands. The adaptive peaks have slid to the east and slid to the west, and the birds have kept flying after them and perching on them again and again. The finches have to stay with their peaks because, as Peter has written, "valleys are steep, that is to say, the intensity of selection is great." The finches have done a lot of flying to stay on their island. *Fortis* has done a lot of evolving just to stay in place.

In these same years the Grants have seen a second oscillation, a change in the fate of the hybrids on the island. The mixed bloods were selected *against* in the first half of the watch, and they were selected *for* in the second half. Up until the flood, a male *fortis* that crossed with a female *fuliginosa* (a small beak) or a *scandens* (a cactus finch) was putting his young at a disadvantage. The hybrids did not prosper. Selection pressure was against intermarriage. But since the flood, selection has reversed. Now a cross with a *fuliginosa* or a *scandens* does the genes of *fortis* a favor.

By putting the two oscillations together, the Grants can begin to understand what is going on with the hybrids. Their misfit data are beginning to fit.

These two oscillations are driven by the same events. They are both governed by the same changes in the adaptive landscape. In an adaptive landscape that is wrinkling and rolling as fast as Daphne, a landscape in which the peaks are in geological upheaval, it can pay to be born different, to carry a beak 3, 4, or 5 millimeters away from the tried and true. Since the super-Niño, some of the old peaks have turned into valleys, and some of the old valleys are peaks. Now a hybrid has a chance of coming down on the summit of a new peak. It can luck onto a piece of the new shifting ground.

In this changing landscape the hybrids may have advantages not only because they are so variable in the dimensions the Grants are measuring. It is also possible that the influx of new genes that is the birthright of the hybrids could translate into a thousand subtle advantages too small for the Grants to measure: benefits that add up to greater physical vigor, even if the bird stays on the same adaptive peak as everybody else. "A hybrid could do all the things others do on an island," Peter muses, "and just be a better piece of machinery generally."

Thoughts like these send Peter back to a paper that two evolutionists, Richard Lewontin and L. C. Birch, published in 1966, "Hybridization as a Source of Variation for Adaptation to New Environments."

We generally think of the adaptive landscape as being more or less fixed and constant, just as we think of the bodies and behaviors of animals as more or less constant. But what happens if the adaptive landscape changes dramatically? What happens, for instance, when a species leaves home and wanders into new territory? Lewontin and Birch suggested in their paper that the genetic changes that accompany the change in range must be "profound," and "if a case could be found of a species rapidly expanding its ecological range, caught *in flagrante delicto*, it might be possible to study the genetic basis of such a change."

Lewontin and Birch found their case in a fruit fly, *Dacus tryoni*, close cousin of the notorious Mediterranean fruit fly, or medfly. *Dacus tryoni* once lived solely on fruits in Australian tropical rain forests. That began to change in the 1850s, while Darwin was writing the book that became the *Origin of Species*. In those same years, farmers "down under" began planting orchards in Queensland. The flies moved out of the rain forest and became a pest in the new apple, pear, and guava orchards. Within a hundred years the flies had expanded their range as far south as Victoria, with sporadic outbreaks in Adelaide, Melbourne, and Gippsland, and in the capital of Australia's Northern Territory, on the Timor Sea, the port called Darwin.

The farther the flies pressed from their original home in the rain forests, the cooler the weather they encountered. In fact their expansion carried them all the way from the tropical zone to the temperate zone. Lewontin and Birch studied historical records and maps and concluded that the flies had been blocked and slowed in their march across the continent chiefly by this change in climate. Laboratory tests confirmed that the strains of flies in the farthest reaches of their range were more resistant to cold than the flies back home in the rain forest, and there was a neat gradation of cold resistance among the flies in between. These were heritable changes, encoded in the flies' genes, and all these adaptations had evolved in this species within a single century.

In terms of the adaptive landscape, *tryoni* was hopping from peak to peak, and as it got farther from the rain forest, each peak was colder and snowier than the one before. Actually the journey was harder on the flies than that, both colder and hotter, since *tryoni*'s migration from the tropics to the temperate zone exposed it to seasonal swings of temperature that were more and more extreme in both directions.

"Such a process of rapid evolution means rapid genetic change," write Lewontin and Birch, "and such change, in turn, demands genetic variation on which natural selection can operate. But where did this genetic variation come from?"

It was possible of course that the variation was already present in *tryoni* back in their primeval rain forest—present in the form of extremely rare genes—and that these genes simply became selected, became more and more common, as the flies hopped from orchard to orchard, farther and farther into the temperate zone. Lewontin and Birch could not rule out that possibility, but they were writing to propose another hypothesis.

Tryoni lives side by side with a second species of flies, *Dacus neohumeralis*. *Tryoni* and *neohumeralis* are sibling species, like Darwin's finches. The flies share most of the same orchards. Mothers in the two species will even lay their eggs inside the very same apple, which means that *tryoni* and *neohumeralis* larvae often grow up side by side, like littermates.

The only thing that seems to hold these flies apart is sex. *Tryoni* copulates around sundown, and *neohumeralis* copulates from mid-morning to mid-afternoon. So the two species are isolated from each other by time though not by space. They look and act so much alike that at least one observer had labeled them as only subspecies. But Lewontin and Birch reason, as the Grants do with Darwin's finches, that "the clear sexual isolation and the maintenance of their separate identities in nature" warrants calling each of them a separate species.

Tryoni has some bright yellow markings, whereas *neohumeralis* is plain dull brown. Lewontin and Birch point out that intermediate forms turn up fairly often in collections: flies with a little mosaic of the yellow and the brown. Careful studies proved that these intermediates are, in fact, what they seem to be: hybrids, products of rare crosses between *tryoni* and *neohumeralis*. So the separation between the species is not absolute, any more than it is among Darwin's finches. One fly likes the lights on, one fly likes the lights off, but every once in a while a pair of them gets together anyway.

"This gene exchange has not been sufficient to merge the species," write Lewontin and Birch, "presumably because of selection against hybrids, but has been sufficient to incorporate foreign species genes into the gene pool of each."

They are two species as closely related as Darwin's finches. They often pass genes back and forth, like the finches. They are held apart by natural selection, as the finches were during the first half of the Grants' study.

With the flies, there seems to be a loose equilibrium in the two gene pools. Alien genes are lost as selection weeds out the hybrids, and

more alien genes spill in again as the rare pair meets and mates somewhere along the invisible border that separates their two kinds.
Lewontin and Birch suggest that it was this introgression of genes that
had led to the rapid adaptation of the flies and allowed them to expand
into a radically new physical and adaptive landscape.

To test this hypothesis, Lewontin and Birch performed an experiment in the laboratory. They collected flies of both species and bred
them in the lab. Then they set up population cages at three
temperatures—20°, 25°, and 31.5° C (68°, 77°, and 89° F)—cool,
warm, and hot for these flies.

Lewontin and Birch allowed populations of each species to evolve
for two years at each of these temperatures. They witnessed the evolution of a new, combined strain that was more fit than either species
is separately. There were marked and rapid genetic changes.

"The introduction of genes from another species can serve as the
raw material for an adaptive evolutionary advance even though the
original hybridization is disadvantageous," Lewontin and Birch conclude. "How often this has happened in nature is another question."

PETER AND ROSEMARY have just seen it happen in nature—on
some of the most remote islands in the world. A new vista is opening
before them. They are looking at a very broad event, whose action has
encompassed the whole of their watch in the Galápagos.

"Under *some* circumstances," Peter says, "the populations are
maintained as separate entities, because any hybridization that occurs,
however rarely, is penalized. The offspring are not as fit. Their chances
of surviving to reproduce are not very good. *Then comes the rare event.*"
A terrible drought, or a plague, or a once-in-a-century flood shakes
up the island, transforms the adaptive landscape so that the peaks and
the valleys are no longer where they were before. The whole adaptive
landscape gets shaken like a rug and thrown down again in haphazard
new wrinkles and folds. Now the birds that fall in what used to be a
valley may find themselves perched on a new, rising peak. Suddenly
they are at an advantage. "That causes a very, very slow fusion of the
populations," Peter says. "That's the *direction* in which hybridization is
pulling.

"But before that goes very far, I think the pendulum will swing
back the other way. And at the very least arrest, and at the most reverse, the process."

"The net result," Rosemary pronounces: "fusion or fission!"

"At times, we think now, hybrids are at a disadvantage," Peter says, "and at times, at an advantage. In the last ten years, the hybrids were at an advantage. But in the ten years before that, the hybrid birds were at a *dis*advantage. So our mental model is one of oscillation, an oscillation of hybrid superiority and inferiority."

They see a sort of vast, invisible pendulum swinging back and forth in Darwin's islands, an oscillation with two phases, each phase lasting a decade or more. "Put the two together, and it is very unlikely that the fusion will go to completion before the wheel of fortune is, so to speak, reversed."

Peter Boag and Peter Grant have projected the consequences of this kind of action for *fortis* and *fuliginosa*. The Grants summarize those results in a paper for the *Proceedings of the Royal Society of London*. "At the observed rate of interbreeding," the Grants write, with "no hybrid advantage and no selection, it would take more than fifty generations, or more than two hundred years, to eliminate the morphological differences between them."

Their estimate is conservative. If they factor in the hybrid advantage that the Grants have seen since the flood, then the change would take less time—somewhere between one hundred and two hundred years. If they factor in the increasing rate of interbreeding, it would take less time yet.

In the two decades they have been watching, the Grants have seen the pendulum swing toward drought, toward flood, and back again. They have seen the adaptive landscape heave in slow motion like whitecaps in an invisible sea. When the landscape returns to something like its condition before the great flood, when the land dries out and the cactus and the *Tribulus* come back into their own, the flow of genes between the species should dry up too. Then, as the Grants write, the hybrids that have flourished in this present interval, in the landscape as it stands now, will be at a disadvantage again, they will be weeded out by natural selection, "and the three species will persist as three separate species, until the next extraordinary El Niño event occurs. Over the past 500 years El Niño events classified as 'strong' have occurred one to three times a century."

If conditions keep oscillating on the islands more or less as they have in the last half of this millennium, there should be no time for the birds to fuse. The very existence and persistence of the thirteen species argue this. "Surely hybridization must have been selected against," Peter says. "So, hybridization has not been strong enough to bring these species into a state of *panmixia*," he says, pronouncing with some pleasure the exotic lilt of the word.

The Grants need to watch longer and study more to be sure. But this is the view that seems to be opening before them. Whenever the adaptive landscape heaves and flings about, like a sea under heavy winds, the hybrids among Darwin's finches will be favored. They will intermingle their genes. But when the landscape returns to the pattern it held before the storm, the birds will settle back to their old peaks, and the sharing of genes will slow again.

THE GRANTS HAVE BEGUN to think about how far all this goes beyond the Galápagos. "Hybridization," as they have written this sabbatical in an article for the journal *Science*, "provides favorable conditions for major and rapid evolution to occur." There are a total of 9,672 species of birds in the world today. Back in 1975 a German ornithologist, W. Meise, estimated that about 2 percent of the younger, more recent species hybridize regularly, and about 3 percent more hybridize occasionally. In 1989 a Russian ornithologist, E. N. Panov, compiled a more extensive list, including every species of bird that has ever been seen, even once, to hybridize. As the Grants note, "No other class of organisms of comparable size is known so comprehensively." And the new numbers look interesting.

The total number of bird species in the world is almost 10,000. Almost 1,000 of them, the Grants write, "are known to have bred in nature with another species and produced hybrid offspring . . . roughly one out of every ten species."

Among some orders of birds the incidence is even higher. Hybridization seems to be quite common among grouse and partridges, also among woodpeckers, hummingbirds, and many species of hawks and herons. It is highest of all among ducks and geese. Of the 161 species of ducks and geese in the world, 67 species have been known to hybridize. In other words, as the Grants note, almost one out of every two species of ducks and geese has been seen to interbreed in the wild.

The actual incidence is likely to be much higher. After all, Darwin's finches have been one of the best-studied groups of birds in the world for most of this century. Yet it is only now, after this extraordinary watch, involving generations of birds and generations of graduate students, that the extent of the mingling of genes among Darwin's finches has come to light. No one has ever followed a set of species of birds in the wild (or any other kind of animal in the wild) with this kind of near omniscience, with every single individual in every generation identified and tracked, its family tree charted and its fate recorded at last in a waterproof notebook with a cross and an R.I.P.

Not so long ago, hybridization among birds was thought to be very rare. In 1965 Ernst Mayr, one of the finest ornithologists and evolutionists of this century, wrote, "On the basis of my examination of random collections, I estimate that perhaps one out of 60,000 wild birds is a hybrid." His estimate may be correct for old and well-established species. But now it seems possible that the interbreeding of birds is more common among younger lineages, where, in Darwin's phrase, we find the manufactory of species still in action. And the process may be important for evolution, the Grants write, "because it produces novel combinations of genes, as well as new alleles [variant forms of the same gene], thereby creating favorable genetic conditions for rapid and major evolutionary change to occur."

It may seem improbable that the crossing of lines could do so much to shape the tree of life. But the power of intercrossing "is not hypothetical," to recycle the phrase with which Darwin introduced the power of natural selection. Among plants the intercrossing of species can create new species, and it can do so literally overnight. "As many as forty percent of plant species may have arisen in this way," the Grants write. That is a huge number of species. Somewhere between a third and a half of all the green things on this earth, and at least half of the world's flowering plants, arrived by the mixing of genes from separate species.

Traditionally, evolutionists have thought of this kind of intermixing and rapid evolution as the more or less exclusive property of the plant kingdom. Mayr concluded that hybridization was unlikely to play much of an evolutionary role among higher animals. Yet that may not be true. Certainly it is rarer among animals than plants, but among birds and many other groups of animals, it seems, hybridization is widespread. It is common in toads of the large genus *Bufo* and in many families of insects. It is extensive among fish, which usually spread their sperm and eggs in the water to be fertilized outside their bodies, rather like plants. Mayr himself cites "occasional or extensive hybridization" among lampreys, trout, salmon, whitefish, catfish, pike, goodeid killifish, live-bearers (including John Endler's guppies), silversides, perch, sunfish, and more.

The flowers we enjoy so much in the plant kingdom are really sperm throwers and sperm catchers. "As we delight in the strange and exotic beauty of orchid flowers," writes a British biologist, "it is salutary to reflect that we are, in essence, looking at their genitalia." Because they are open to the winds, they catch a lot of alien sperm that animals' sperm catchers are more likely to dodge. Being animals we find the flowers' arrangement peculiar. But in the big picture, in the way our lines grow and split on the tree of life, our kingdom may not be so different from theirs. "Animal species may be more like plants than is generally realized," the Grants write. Animals may mingle their genes almost as freely as the trees and flowers that send out their sperm to drift on every breeze, and open their flowers to catch the sperm from every breeze. Many animals may have their "genetic systems open to invasion, especially early in their existence as quasi-independent lineages."

For plants the advantage of all this interbreeding is obvious. Mayr has put it succinctly: "Plants cannot move. A seed germinates where it drops and must succeed or die." So as the plants' pollen is swept from one plant to another by winds and insects, hybridization is not only inevitable but also desirable, because so many myriads of seeds will fall and sprout in adaptive landscapes that are different from those of their parents. Here natural selection favors great genetic variability, and hybridization is one way to generate it fast. Cross a tree with star-shaped leaves and a tree with spear-shaped leaves, and you can get generations of hybrid leaves with shapes like splayed hands, pyramids, hearts, and arrowheads. That is only the variation that catches the eye—imagine the variation beneath the surface.

Because the Grants are watching so closely, they can see that even on the same desert island, on a lump of rock that looks to the casual eye as changeless as the moon, the adaptive landscape varies with extraordinary energy from decade to decade. So the birds that are bound to this little island, breeding where their line has bred for generations, may often need as great an infusion of fresh variation as plants whose seeds drift hundreds of miles on the wind.

TO THE GRANTS, the whole tree of life now looks different from a year ago. The set of young twigs and shoots they study seems to be growing together in some seasons, apart in others. The same forces that created these lines are moving them toward fusion and then back toward fission.

The Grants are looking at a pattern that was once dismissed as insignificant in the tree of life. The pattern is known as reticulate evolution, from the Latin *reticulum*, diminutive for net. The finches' lines are not so much lines or branches at all. They are more like twiggy thickets, full of little networks and delicate webbings. This sort of reticulate evolution doesn't bind lineages together forever; eventually they part ways or fuse. But it may be a general and hitherto neglected feature of the origin of species.

Ever since the Grants and their team published the first news of hybrids among Darwin's finches, evolutionists have been talking and writing about this new view the finches are helping to open up, the implications of a reticulate tree of life.

"Instead of thinking of the evolutionary chart of the finches as a well-developed family tree with clean branches heading off in distinct directions," writes the evolutionist David Steadman, who is an authority on the fossil remains of Darwin's finches, "I find it useful to think of it as a young bush in which branches are so tangled, untrimmed, and interrelated that evolutionary directions remain jumbled and tentative." He writes, "It is as if, like young adults, they are experimenting with different adult identities, some aspects of which they will keep and some they will discard."

"In the short term," writes another evolutionist, Jeremy Searle, "they are not following entirely independent evolutionary pathways." A novel gene that evolves in one species can spread to others. Life would be so much simpler if lines of animals would only keep to themselves, Searle writes, only half-jokingly. That should not be too much to ask: it is the zoologist's standard working criterion of a good species. But "things are not so easy for zoologists," Searle concludes. "It is disappointing that even Darwin's finches do not seem to quite fit the bill."

A third evolutionist, Robert Holt, is also greatly struck by this blending of competing lines. "Species that are competitors over ecological time," Holt writes, "may be mutualists over evolutionary time, each providing a store of genetic variation that can be tapped by the other.

"Maybe we should all be grateful that Mother Nature is a bit slovenly when it comes to reproduction, for this may ultimately permit the unfolding of the bountiful diversity of life on Earth."

The old vision of the tree of life was plain, neat, stark; this view is softer, messier, more tangled, and more alive. In a way it is also more sympathetic. Clearly the lineages of Darwin's finches do compete: they struggle and push one another apart according to Darwin's principle of divergence; they play endless games of King of the Mountain. But at the same time the birds on their separate islands and lonely peaks are not as solitary as they had seemed. They are full of fissions and fusions, competition and cooperation, like brothers and sisters in a nuclear family, bound by a thousand nuclear ties and tensions; or like the old royal families of Europe, exchanging princes and princesses to link their lines. The birds pass invisible messages back and forth, swapping genes as casually as good neighbors exchange recipes, tools, or limericks. They are secret sharers, communing on their long voyage, open to suggestions. Their lines come together and come apart, and in this way the birds are created and re-created, again and again.

The apparent fixity of species once seemed the greatest argument against evolution, just as the apparent fixity of the earth once seemed a commonsense argument against Copernicanism. Now the satisfying and reassuring sameness that once encouraged Aesop and other fable spinners to speak of The Fox, The Owl, The Wolf, The Whale, and The Crow seems more illusory than ever before. "All is flux," said the Greek philosopher Heraclitus; "everything flows." The forms and instincts of living things, the invisible borders among them, and the very coasts and landscapes they inhabit are all more fluid and in more flux than even Heraclitus could have imagined.

Chapter 14

New Beings

> Hence, both in space and time, we seem to be brought somewhat near to that great fact—that mystery of mysteries—the first appearance of new beings on this earth.
>
> — CHARLES DARWIN,
> *Journal of Researches*

A Victorian gentleman contemplates the skeleton of a stork. He has a yardstick, a divider, a set of calipers, a stack of notes. With a cloth ruler stretched between his hands and a pencil clamped in his mouth, like a tailor, he stares up in the general direction of the bird's beak as if to say, "What do I do now?"

This portrait of the naturalist at work was painted in 1879 by Henry Stacy Marks, a member of the Royal Academy of Arts. Marks called it *Science Is Measurement*.

"When you can measure what you are speaking about and express it in numbers you know something about it," declared the Victorian physicist Lord Kelvin, in 1883, "but when you cannot measure it, when you cannot express it in numbers, your knowledge is of a meagre and unsatisfactory kind; it may be the beginning of knowledge, but you have scarcely, in your thoughts, advanced to the stage of *science*, whatever the matter may be."

The painting and the Kelvin quotation are reproduced together on the frontispiece of *Album of Science: The Nineteenth Century*, a book of pictures edited by the historian L. Pearce Williams. This sabbatical, one of the Grants' many friends at Princeton brought them a copy of the book; opening its covers to the frontispiece gave Peter a moment of high glee. He read the Kelvin quotation aloud to Rosemary.

"Your knowledge is of a *meagre* and *unsatisfactory* kind!" Rosemary repeated, in the voice of a Scottish hanging judge.

Science Is Measurement,
by Henry Stacy Marks.
The London *Graphic*, 1879

"That makes one look pretty small, doesn't it!" cried Peter.
"Oh, that's lovely!"

"*This* is interesting," Peter said, and read the historian's commentary:

> The painting illustrates the puzzlement of the natural scientist,
> whose subject offers little opportunity for significant measurement.
> He was faced with the question of whether all of natural history was
> to be excluded from "science"—or whether the close observer of
> nature was as near to scientific "truth" as the mathematical analyst.

"Yes," Peter said. "I think for much of this century too there was
a prejudice against what we now call ecology, on the grounds that you
can't measure anything with precision—and if you can it's probably
not very interesting.

"That was a prejudice among people who *could do* very precise
measurements in laboratories—physicists and physiologists and so on."

The Grants' measurements make a wonderful reply to Marks's

painting, and to the doubts of Darwin's century. They are as quantitative and rigorous as even Lord Kelvin would have wished. While working on their hybrids this sabbatical, for instance, the Grants extracted a small sample from their shelves of data and carried out a test of the predictive power of Darwinian theory. They took the mean length, width, and depth of *scandens* beaks and *fortis* beaks on Daphne in the year 1984. They inserted each of these numbers into a succinct mathematical formula, together with an assortment of key variables (including the separate heritabilities of beak depth, width, and length; the ways each of these characters affects the others; the ways seed abundances on the island went up or down between 1984 and 1987; and the separate ways that seed abundances affect beak depth, width, and length). Then they let the formula predict the mean length, width, and depth of *scandens* beaks and *fortis* beaks on Daphne in 1987. Finally they compared the results their equations predicted with the actual results of Darwinian evolution on Daphne between 1984 and 1987.

Mean *fortis* beak width, for instance, was 8.86 millimeters in 1984. The formula predicted that by 1987, give or take a minute fraction of a millimeter, the mean width should have dropped to 8.74 millimeters. The actual width of *fortis* beaks on Daphne in 1987 was 8.74 millimeters.

They were "spot-on," as Peter says. Every single one of the numbers was right.

"Those who criticize the theory of evolution by natural selection often do so on the grounds that it is impossible to test it by making quantitative predictions," the evolutionary biologist Jeremy J. D. Greenwood notes in a recent commentary in *Nature*. "Rosemary and Peter Grant now show this view to be wrong. . . . Their predictions have been precisely correct."

If the Grants' measurements were not as precise as this, they would have missed virtually all of the action they have seen in the Galápagos in the last twenty years. Certainly they never would have caught a glimpse of the invisible pendulum that they now see swinging across the decades in the Galápagos. "We only discovered that by doing an analysis of the numbers," Peter says. "It doesn't shriek out from the data. It's a subtle matter of probabilities shifting by a few percent."

Their discovery of the rise and fall of the mixed breeds takes nothing away from the power of natural selection. On the contrary. It shows more vividly than ever before that Darwin's finches are new beings on this earth. The intense selection pressures that shape and reshape the beaks of the finches also keep all these beaks from disappearing. Darwin's process created the many out of one, and Darwin's process is in the act of their creation even now. If natural selection did not go on working hard on each island, in each generation, the many would very soon vanish into one again.

Darwin's finches are not like Michelangelo's Adam, who raises his finger languidly to meet the down-stretched finger of God: the first man, molded of clay, half-raised from earth, created in an instant. These birds are more like Michelangelo's *Prisoners*, the famous statues he left half in and half out of the marble, so that looking at them today we can almost see and hear the sculptor's chisel at work. The birds are alive and breathing, but they are unfinished; in the Galápagos the sculptor is still at work, measurably and demonstrably. The Grants' discovery this sabbatical makes the action of the chisel only more dramatic to contemplate. The more the birds are able to fuse, the more impressive the work of the sculptor: the faster the chisel must be flying, to keep them all apart, as if it were not sculpting in stone at all, but writing in water. Roughly one out of ten of the finches born on the desert islet of Daphne Major now are hybrids, and the hybrids are doing better than any of the others on the island. In a blink of evolutionary time, all of Darwin's finches could run together and congeal, and the sculptor's art would be lost. As the evolutionist Ernst Mayr once pointed out, the tendency toward fusion, the "successful leakage of genes from one species into another," is "a self-accelerating process." Each case of introgression weakens the invisible barriers between two species and leads to increased frequency of hybridization, a process that if unchecked will spill downhill faster and faster, "until ultimately the two species are connected by a continuous hybrid swarm."

The finches are not yet carved completely apart from one another or from the ancestral stock, from the line of birds that first colonized the islands millions of years ago. If the chisel were not flying fast, the work of the carver would soon disappear without a trace, fusing back into the block of the living stone.

This same tension between fission and fusion runs through all the kingdoms of animals and plants. Everywhere hybrid swarms are rare; good, solid, more-or-less-separate species are common. Yet in many of the birds flying overhead, in many of the fish in the sea, and in almost

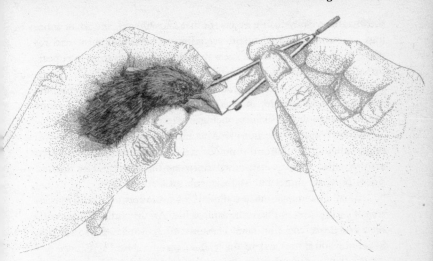

Measuring the length of a beak.
Drawing by Thalia Grant

all of the green things growing around us the genes are intermingling. The chisel is hard at work daily and hourly in every landscape on the planet.

WHY ARE THERE SO MANY KINDS of animals? Adaptive radiations like Darwin's finches are the essence of the answer. They are in progress all over the Galápagos (the mockingbirds, the cacti, the sharks, the tortoises, the torchwood) and all over the planet. In the Hawaiian Islands, the lineage of a single finch has radiated into more than forty species, with forty beaks, including seed crushers, bug catchers, nectar sippers. Their beaks are even more divergent than the beaks of Darwin's finches. The *akiapolaau* has one of the strangest. It strips bark with its lower mandible, like a knife; then it pries out bugs by poking the wood with its very long, thin, overgrown upper mandible, like a needle. Two tools in one, a double-bladed pocketknife.

Also in Hawaii, a few lost fruit flies that blew in over the Pacific millions of years ago have now radiated into somewhere between five hundred and one thousand species. More than a third of all the species of fruit flies on earth belong to this single adaptive radiation. There are predatory flies and parasitic flies; nectivorous, detritivorous, and herbivorous flies. Some of them are the size of pinheads, and some are as long as a child's thumb; a few, according to an evolutionist who has stared at them face to face, "have strange broad heads with eyes positioned far apart like those of hammerhead sharks."

In the Great Lakes of East Africa a group of cichlid fish has achieved a whole series of these starbursts. In Lake Victoria alone about two hundred species of cichlids have evolved from a single ancestral stock within the last 750,000 years. Some feed in the water column, some at the bottom; some eat snails, others eat fish, still others eat fish scales. One species, more gruesome than the vampire finch of the Galápagos, plucks out and eats fish eyes.

All of these radiations are displays of evolution in action, success stories in the contemporary history of life. Any evolutionist can rattle off half a dozen more of them: Hawaiian silverswords, leiognathid fish, heliconid butterflies, and so on, and so on.

At every moment in the history of life, including our own moment, adaptive radiations like these are in progress all over the earth. They decorate the map of every part of the globe in every era like the compass roses of the old cartographers, or cross sections through the branches of a growing tree.

The history of these adaptive radiations is the history of life, from the explosive radiations of the bizarre fauna of the Cambrian, 540 million years ago, to the radiation of the first jawless vertebrates, the Agnatha, in the Ordovician, 500 million years ago; the radiation of fish in the Devonian; amphibians and insects in the Carboniferous; dinosaurs and mammals, beginning in the Triassic; angiosperms and yet more insects in the Cretaceous; and in the Pleistocene, a few million years ago, radiations of herbs and human beings.

"You had better stay with Darwin's finches or else there is no end to your labors," Peter says. "Stay with Darwin's finches or the finches will seem at last like a little momentary excitement in your youth."

THE GRANTS' SABBATICAL YEAR is already over; sabbaticals never last as long as one would like. But Rosemary and Peter are trying to pretend that this one is still going on. They are sitting together over a brown-bag lunch in a corner of Peter's office, wedged between classes and seminars, planning their next trip to the islands and puzzling together about the origin of species.

With nearly 99 percent of their finches, the As still mate As, and the Bs mate Bs. "They are definitely species," he says.

"They're undoubtedly species," says Rosemary. "They differ in song, size, and shape. It's easy for us to tell them apart, and they tell each other apart."

And yet, when Peter and Rosemary consider the rise of the misfits, the 1 or 2 percent of As that choose Bs and Bs that choose As, the finches do not look the same as they did one year ago.

"The most intriguing feature," says Peter, "is the possibility of speciation. By putting genes together like this, you make combinations that could take off, that could be the starting point of a new evolutionary direction not easily within reach of either species."

Evolutionists have talked about this notion for decades, but it never seemed as compelling to Peter and Rosemary as it does after this sabbatical. They see now how much mixing is going on all the time on Darwin's islands, and how often fate seems to smile on the oddballs, which to their experienced eyes are very strange birds and queer ducks indeed. The Grants are wondering whether some of these mixed-bloods on Daphne Major could be "potential escapees," in Peter's phrase, from the constraints that bind the genes of their kind; whether birds like these could be the beginning of a really new departure, "the starting point of a new evolutionary lineage."

"An evolutionary response *could* follow," says Rosemary. A new variant among the broods of novelties could go off along a new road. But the window of opportunity might be very small, she thinks. "There'd have to be very strong selection." Without a selection pressure that favored a novelty-shop beak, promiscuous interbreeding would soon erase it, she says, "because you'd constantly get backcrossing."

"Undoubtedly it's a complicated matter," says Peter. Not for the first time, he begins sketching the future of a lucky hybrid line on his paper napkin. Rosemary jumps up and heads for the blackboard, crying, "Let's see if . . ."—if their visions match.

"If you start off with hybrids, they're at a terrific numerical disadvantage," Peter says, drawing a few scattered dots between two thick clouds, representing As and Bs. In his sketch, the hybrids are floating here and there in the space between the two clouds, and they are vastly outnumbered. Because they are in a minority, they are more likely to mate with one of the old lines than to mate with another hybrid. "So their offspring are halfway back. A dilution of novelty." He adds an X-axis and a Y-axis to the graph, plotting beak length against beak depth.

Rosemary stops sketching on the blackboard, darts back to the table to see what Peter is doodling on his napkin, then adds a few finishing touches at the blackboard. She and Peter are both thinking the same way. She draws a long thin arrow that shoots off between her clouds of As and Bs. That is the narrow path an evolutionary novelty might follow to escape in a new direction.

"A dense cloud of birds," Peter says playfully, studying Rosemary's chalk drawing. "It's all preliminary. . . ."

"Could be wrong," says Rosemary.

"Speculative," says Peter.

"Much better off with numbers," says Rosemary.

"We have ourselves at the forefront of pure ignorance."

"We've got to do more reading," Rosemary concludes. "Reading, thinking. . . ."

"Measuring," says Peter.

PART THREE

G.O.D.

The river is moving.
The blackbird must be flying.

— WALLACE STEVENS,
"Thirteen Ways of Looking
at a Blackbird"

Chapter 15

Invisible Characters

> Immediately the fingers of a man's hand appeared and wrote on the plaster of the wall of the king's palace, opposite the lampstand; and the king saw the hand as it wrote. . . . The king said to the wise men of Babylon, "Whoever reads this writing, and shows me its interpretation, shall be clothed with purple, and have a chain of gold about his neck, and shall be the third ruler in the kingdom."
>
> —*Daniel 5:5*

The Grants have one more Galápagos archive, which they keep not far from their offices in Eno Hall; and this is the one that would really have astonished Darwin. The shortest way there is through Princeton's Natural History Museum, past the black-boned *Allosaurus*, the extinct Irish Elk, the remains of a Neanderthal from France, and an early *Homo sapiens* from Israel; past broken jaws and speechless fragments, including bits of *Eohippus*, *Pliohippus*, *Dinohippus*, and *Equus*, the sequence of fossils that Huxley used more than a century ago in his lecture "The Demonstrative Evidence of Evolution."

One floor below the museum there is a long drab basement corridor. Toward the far end, a steel stairway drops sharply as if into the lower level of a ship. This stairway leads to a subbasement, or rather a sub-subbasement: C-Level of the building next door to the museum, the George M. Moffett Biological Laboratory.

Moffett C-Level throbs like an engine room. The din of compressors, fans, and generators; the drone of fluorescent tubes; and the narrowness of the passageway combine to give the place a bowels-of-the-beast feeling. The air reeks of formaldehyde and chemicals less familiar.

This entrance to C-Level is constricted by refrigerators marked

CAUTION—RADIOACTIVE and by an oak natural-history cabinet of the kind that was popular in Darwin's century, full of dissections made by long-dead Princeton biology students: Head of a Rattlesnake, Lungs of a Hawk, Stomach of a Screech Owl, Skeleton of a Bat. The cabinet is topped with baseball trophies of young men waiting for the pitch.

Down the corridor, just past an emergency eye-wash station, and an emergency shower, the way is half-blocked by a freezer chest. Open its lid, and clouds of vapor spill out onto the linoleum tiles of the sub-subbasement.

At the bottom of the chest, stacked in hundreds of plastic vials, are the samples that Rosemary and Peter are collecting drop by drop in the Galápagos: the blood of Darwin's finches.

"You have most cleverly hit on one point, which has greatly troubled me," Darwin wrote Huxley, after Huxley had read the *Origin*; ". . . what the devil determines each particular variation? What makes a tuft of feathers come on a cock's head, or moss on a moss-rose?" Darwin never did find the answer. He was convinced that variations are the origin of species, but he did not know the origin of variations. "Our ignorance of the laws of variation is profound," he confesses in the *Origin*. "Not in one case out of a hundred can we pretend to assign any reason why this or that part has varied."

Darwin did predict that the reason would be discovered. Some kind of secret writing would one day be detected and deciphered in the bodies of living things. He pictured this code as a swarm of letters streaming through the blood, invisible characters that meet and unite in each fertilized egg: "and these characters, like those written on paper with invisible ink," he wrote, "lie ready to be evolved whenever the organisation is disturbed by certain known or unknown conditions."

Like Belshazzar, king of Babylon, who saw the writing on the wall, Darwin knew the characters in the blood would turn out to be of the utmost significance. But Darwin could not see the writing himself, and he had no Daniel to read it for him.

Today, biologists call Darwin's invisible characters "genes," from the Greek verb meaning "give birth to," the same root as genius and generation; and they can read the code in the blood.

"FIDDLY WORK TO DO in this kind of weather," says Peter Boag. "Not very much fun. It's a very finicky technique."

He stands at his laboratory bench on a muggy afternoon in August, fussing with a small plastic bag. He has already extracted molecules of deoxyribonucleic acid, DNA, from a drop of the frozen finch blood. He has bathed the DNA in a solution of enzymes, which chopped the DNA into millions of fragments, snipping only at selected points, like manic but intelligent scissors. Boag has sorted these snippets in a sieve made of electrically charged gelatin, and transferred them all onto a swatch of nylon. Now he has the nylon floating in a Seal-A-Meal bag. He is trying to get the bubbles out of the bag so that he can seal it up. Tiny bubbles keep clinging to the nylon.

The liquid in the bag looks like plain water, but it is loaded with radioactive phosphorus, P^{32}. A transparent screen like a TelePrompTer stands between Boag and the lab bench, protecting his chest from the radiation. "When in water the nylon's reasonably well shielded," he says. "But we do the fiddly parts behind the screen." He wears the same gloomy look with which he once knelt in the hot dust and blazing sun of Daphne Major, sorting seeds of *Portulaca*, *Cacabus*, and *Heliotropium*.

For Boag the drought on Daphne Major took place in another lifetime. He and Laurene Ratcliffe are both professors now at Queens University in Kingston, Ontario. They have a three-bedroom suburban house near the university, one slightly cranky car, three small children, a black Labrador retriever, and a Siamese cat. Boag watches Galápagos evolution now from this laboratory bench.

He and the other alumni of the Finch Unit are among the first cohorts of evolutionists to come of age after Watson, Crick, and the revolution in molecular biology. Every year, thanks to this revolution, more and more startling manipulations of DNA become possible—not only possible but routine. Boag went molecular seven years ago, and so have many other former watchers in the Finch Unit, although Laurene has so far cheerfully abstained, and so have the Grants.

("It really is a foreign language," Peter Grant says. "I suppose ours is too, but I do have the impression that theirs is harder.")

Boag swishes the nylon around in the bag, seals it up, and sets it aside to sit overnight in its radioactive bath. The DNA fragments on the nylon are still invisible. Tonight a chosen few among them will turn hot. In the jargon, these fragments will have been tagged by the P^{32} probe. Tomorrow, he will press the nylon against a sheet of X-ray film, and keep it there long enough for the hot spots of DNA to expose the film and make a picture.

Using that X-ray as a guide, Boag can select a single fragment of special interest, multiply it a millionfold, and make another, much more detailed X-ray.

"When that picture comes out of the film developer," he says, "you're the first one who's ever looked at the DNA sequence of one of Darwin's finches."

He holds up a sheet of X-ray film, one of dozens he and his students have already made. The film is covered with rows upon rows of little ghost-gray blobs.

"This is a cactus finch, *scandens*," he says.

BOAG GETS SOME OF HIS BLOOD SAMPLES from the Grants, and some he collected himself on a brief trip to the Galápagos in 1988. The X-ray he is holding up is a portrait of a single gene from a cactus finch that he caught, bled, and freed again on the island of Santa Cruz: "Just a random individual from Academy Bay."

The gray smudges in the X-ray are aligned in four columns labeled G, A, T, and C. Boag scans these columns familiarly, reading from the bottom to the top. A Galápagos finch has about one hundred thousand genes, roughly the same number as a human being. The genes are spelled out in a total of about one billion letters, an average of ten thousand letters to a gene. The story is big but the alphabet is small: there are only four letters, named for the four chemical compounds that, as Watson and Crick discovered, make the treads in every spiral staircase of DNA, rather like the leaden letters in a printing press. Their chemical names are guanine, adenine, thymine, and cytosine: G, A, T, and C.

The gene in this X-ray comes from the mitochondria of the cactus finch. Mitochondria are the dark, peppery granules in which each cell converts oxygen into energy. This particular gene codes for the enzyme cytochrome *b*, which plays a role in the oxygen-conversion process. For the past few years Boag has been isolating this same piece of the cytochrome *b* gene in species after species of Darwin's finches, working with a Swedish postdoctoral student, Hans Gelter, of the University of Uppsala.

If species were created once and for all, if they all came tumbling to life finished and polished, as Milton paints them in *Paradise Lost*, then each species of Darwin's finches would possess a fixed, permanent, never-changing set of genes. But genes are not fixed. The one hundred thousand genes of Darwin's finches are shuffled and cut in every generation, like a mammoth deck of cards. Each finch egg in each cactus tree receives a unique, absolutely unprecedented combination of genes. That is why every finch in the Galápagos is endowed with its own set of measurements of beak, wing, tarsus, and hallux, to say nothing of thousands of variant forms of submicroscopic enzymes like cytochrome *b*.

A finch as it flips pebbles looking for *Tribulus* seeds is bombarded by cosmic rays from outer space, by ultraviolet radiation from the sun, by miscellaneous lost wandering molecules that thonk into its DNA strands like loose cannons, even by the thermal motion of its own atoms and molecules: the thousand natural shocks that flesh is heir to. These aerial and internal bombardments agitate the billion letters of its DNA. Every twenty-four hours, inside every living cell, about one hundred copies of the letter C are half-sprung from their spiral railings. Squadrons of enzymes travel day and night along the strands of DNA and mend the broken Cs. In effect these enzymes are proofreaders. Sometimes they put a C back wrong, and forever after that particular C reads G.

Proofreading mistakes introduce haphazard new variants into the finches' DNA, adding, subtracting, and switching letters around. If a mutation occurs in the DNA of an ordinary cell, it can cause cancer. If it occurs in a sperm or an egg cell, it can be passed on to the next generation. If it is more or less neutral, or if it happens to confer a slight benefit, its bearers may survive and pass it on again and again.

In Boag's X-rays he finds slight differences in the sequence of letters between one finch and another. The X-ray he is holding in his hand, for instance, shows exactly three hundred letters in the cytochrome *b* gene from his cactus finch. This sequence does not quite match the corresponding three hundred letters in a tree finch. Three out of the three hundred letters are different.

Ground finches and tree finches are closely related, since they arose on Darwin's islands quite recently in evolutionary time. All of the Galápagos finches are close-set twigs on the same small branch of the tree. It makes sense that about 99 percent of their invisible characters should be exactly the same. None of these lines have had the time to accumulate many new mutations.

Birds on other, more distant branches of the tree of life carry many more differences in their DNA: a few percent. The farther apart two species sit in the tree, the more differences there are in their DNA.

Small tree finches. From Charles
Darwin, *The Zoology of the Voyage of
H.M.S. Beagle.*
The Smithsonian Institution

But every living thing carries its code in the same invisible
characters, always the same four letters, because ultimately every living
thing on earth shares the same ancestor, about four billion years back,
near the very birth of the planet.

By looking at DNA sequences, evolutionists are now gleaning
more and more of the lost history of life. They are filling in the family
tree from the youngest twigs at the top to the oldest forks at the bottom. Investigators looking at DNA have discovered, for instance, that
crows evolved in Australia, that storks are close kindred to vultures,
and that mushrooms and toadstools are more closely related to animals
than to plants. Evolutionists can learn these sorts of genealogical secrets from mutations in DNA precisely the same way historians learn
from spelling errors in old manuscripts.

Historians are aware, for example, that Darwin's personal spelling
habits evolved during the course of the voyage of the *Beagle.* He was
sending his journal home piecemeal and getting letters back from his
family. On his twenty-fifth birthday, February 12, 1834, his sister Susan,

whom he called "Granny," wrote to him praising his journal: "and what a nice amusing book of travels it w^d. make if printed," she wrote; "but there is one part of your Journal as your Granny I shall take in hand namely several little errors in orthography of which I shall send you a list that you may profit by my lectures tho' the world is between us.
—so here goes.—

WRONG	RIGHT ACCORDING TO SENSE.
loose. lanscape. higest	*lose. landscape. highest.*
profil. cannabal	*profile. cannibal.*
peacible. quarrell	*peaceable. quarrel.*

—I daresay these errors are the effect of haste, but as your Granny it is my duty to point them out."

A year and a half later, after his stop in the Galápagos, Darwin's sister sent him another spelling lesson: "I cannot think how you c^d. write such a collected account of your travels when you were Galloping so many miles every day.—When I have corrected the spelling it will be perfect, for instance *Ton* not *Tun*, *lose* instead of *loose*.—You see I am still your Granny."

The historian of science Frank J. Sulloway has now compiled a table of all of Darwin's spelling habits during the voyage. (Granny's nightmare!) Sulloway collated every variant spelling in the more than three thousand manuscript pages that Darwin had scribbled aboard ship. By noting the dates at which each of these errors drifted in and out of Darwin's letters, journals, and field notes, the historian was able to zero in on the date when Darwin, sitting in his cabin, inspected the Galápagos mockingbirds and realized that they might "undermine the stability of Species."

Five words helped Sulloway pin down that fateful moment. There was a stretch toward the end of the voyage when Darwin often spelled *occasion, occasional,* and *occasionally* with a double *s*, and *coral* with a double *l*. He also sometimes misspelled the ocean he was floating on—the *Pacifick*. Because Darwin's ornithological notes contain *occassion, corall,* and *Pacifick*, they had to have been written after November 1835 and before mid-September 1836. (From watermarks and other evidence, Sulloway narrows the date to June or July 1836.)

In much the same way, Boag and his postdoc Hans Gelter are trying to sort out how Darwin's finches have branched since the birds first arrived in the islands. Lack drew his chart of their family tree by comparing and contrasting the finches' sizes, shapes, feathers, and beaks—especially their beaks. Lack assumed that the most distinctive forms among the birds had been diverging for the longest time. By this reasoning the warbler finch (which looks so little like a finch that Darwin mistook it for a true warbler) was the first species to split from the ancestral stock. Then the tree finches split from the ground finches. Then the tree finch line and the ground finch line each put out branches and twigs of their own.

Boag and Gelter are preparing the equivalent of Sulloway's table of spelling errors. A mutation in cytochrome b that appears in tree finches, but not in any of the six species of ground finches, is likely to have arisen after the tree finches and ground finches diverged and went their separate ways. A mutation that appears in the small, me-

Warbler finches. From
Charles Darwin, *The
Zoology of the Voyage of
H.M.S. Beagle.*
The Smithsonian Institution

dium, and large ground finches, but in none of the other species of Galápagos finches, is likely to have arisen after those three diverged from the rest. Given a large enough table of data, a computer program can crunch the numbers, weigh the probabilities, and draw a most-likely phylogenetic tree of Darwin's finches.

No one knows which species was the first to reach the islands— the founder of all these lines, the bird to put at the trunk of the tree. Over the years evolutionists have proposed many candidates for the honor. At the moment several species of continental birds are in the running, including *Melanospiza richardsonii*, which now lives only on a single island in the West Indies, but may once have had a much wider range; and the blue-black grassquit, *Volatinia jacarina*, a finch that is common throughout Central and South America, including the whole length of the Pacific coast. Boag hopes that by looking at the DNA of all these birds he can help sort out which is most closely related to Darwin's finches. Which one, for instance, carries a gene for cytochrome *b* that most closely resembles the finches in the Galápagos?

Boag has a reputation in the Finch Unit as the most careful, cautious, detail-minded watcher in the history of Daphne Major. He and Gelter have been working at their project for several years and many rolls of Seal-A-Meal bags.

But now there is an untidy complication, because genes are not fixed and permanent. They are manuscripts that are still being written, and among Darwin's finches the rate of writing and revision is accelerating.

"WE HAVE GOOD REASON to believe," Darwin wrote, when the secret of variation was still a tightly lidded black box, ". . . that changes in the conditions of life give a tendency to increased variability."

Studies of DNA in laboratory cultures of bacteria have borne out this point and thrown light on its causes. In times of stress, when the temperature shoots up or down, for instance, or the environment goes suddenly more wet or dry, colonies of bacterial cells in a Petri dish will begin to mutate wildly. This is known as the SOS response, for the international distress signal Save Our Souls, Save Our Ship. It increases the chance that at least a few of the cells in the Petri dish will survive the disaster of the new conditions.

The SOS response has been observed in the DNA of maize when it is shocked by hot or cold temperatures. Recently it has been discovered in yeast. Apparently many different kinds of living cells can switch up their mutation rate under stress and relax it again when the stress dies down. They can also make select portions of their DNA unstable in the extreme. It is almost as if cells keep a lid on their own evolution—and stress blows the lid.

When stressed in a Petri dish, many cells of *E. coli*, whose normal habitat is the human gut, will even open pores in their membranes and take in DNA from outside their cell walls. Strands of naked DNA are always floating around these bacterial colonies like scraps of old newspaper. The living cells open up, take in the naked DNA, and patch some of it into their own genes. The process is known as transformation, and it can be stimulated by the stress of unfriendly chemicals or ultraviolet radiation.

"The implications . . . are significant," writes one molecular evolutionist, John F. McDonald; "at precisely those challenging moments in evolutionary history when major adaptive shifts are required, genetic mechanisms exist that increase the probability that the appropriate variants will be provided."

What the Grants are seeing now on Daphne implies that something like this is going on at this moment among Darwin's finches. If the finches always bred like with like, they would keep their gene pools separate. Their stores of variation, their unique spellings of thousands upon thousands of genes, would be kept distinct. Each species would have its own set.

But these gene pools are far from separate. Genes are continually flowing back and forth between them, by the crossing of species and the backcrossing of hybrids. The islands are a melting pot.

If they never intercrossed, the birds would be much more stable and uniform in their DNA and in their beaks and bodies. Intercrossing raises their variability. In fact, this appears to be the secret of the remarkable variability of Darwin's finches. The Grants have calculated that if only a single immigrant on Daphne were to inject its genes into each species of finch in each generation, that infusion of fresh genes would suffice to keep all of the finches' gene pools from running low. One injection per species per generation would be enough to keep the birds as variable as they were when the Grants began watching Daphne in the 1970s.

But now the flood of fresh genes is pouring in at many times that rate, and the rate is still rising. The triumph of the mixed-bloods on Daphne implies that the DNA of Darwin's finches is becoming more rich and strange than anyone ever imagined. The hybrids are passing their motley collections of alien genes throughout the quasi-separate species on Daphne. It is as if the finch watchers were collecting the manuscripts of not one but thirteen young voyagers, all of whom are still writing and still learning to spell, and many of whom are copying from each other, with sentences and even whole paragraphs and pages flying back and forth.

All this flying DNA is rapidly raising the finches' variability, like the SOS response of *E. coli* in a Petri dish. It is even possible that the influx of new genes is increasing the birds' mutation rates. Such an effect has been observed in laboratories, in the DNA of hybridizing *Drosophila*.

Times are changing: the birds are under stress, and they are eroding and breaking down their invisible coasts, taking in new genes, growing more variable. The flood of 1982 has set off a flood of genes, a sharing of secret messages, a flying shuttle of invisible characters, a wild mixing party, a huge upsurge in the amount of hidden variation in Darwin's finches.

When the first finches alighted in the Galápagos millions of years ago, the stress of settling the young volcanoes must have been extreme. The intensity of natural selection, the rate of interbreeding, and the speed of evolution must all have been extraordinary. Now something rather like that seems to be happening again.

Chapter 16

The Gigantic Experiment

Man therefore may be said to have been trying an experiment on a gigantic scale; and it is an experiment which nature during the long lapse of time has incessantly tried.

— CHARLES DARWIN,
The Variation of Animals and Plants under Domestication

From his first hour on San Cristóbal, when he shoved a hawk off a branch with the muzzle of his gun, and saw a finch swatted dead with a hat, Darwin knew that the birds of the Galápagos had no experience with human beings. When he mentions the Galápagos birds in his journal, it is almost always to exclaim at their appalling innocence.

"I saw a boy sitting by a well with a switch in his hand, with which he killed the doves and finches as they came to drink," Darwin writes, after visiting the prison colony on Floreana. "He had already procured a little heap of them for his dinner; and he said he had constantly been in the habit of waiting there for the same purpose."

"One day," Darwin writes, "a mocking-bird alighted on the edge of a pitcher (made of the shell of a tortoise), which I held in my hand whilst lying down. It began very quietly to sip the water, and allowed me to lift it with the vessel from the ground. I often tried, and very nearly succeeded, in catching these birds by their legs."

Darwin had been reading the second volume of Lyell's *Principles of Geology*, which he received in his mail at the port of Montevideo. The whole volume is an argument against the possibility of evolution. But Lyell does write about the changes that an invader might bring to an island. When the first polar bears landed in Iceland on an iceberg, for instance, the chaos must have been "terrific," Lyell argues. The bears must have feasted on the deer, the foxes, the seals, even some of the birds. With fewer deer on the island, the local plants would have prospered, and also the insects that fed on the plants. With fewer foxes, the ducks would have multiplied, thinning the schools of fish. In this way "the settling of one new species," Lyell wrote, might alter "the numerical proportions of a great number of the inhabitants, both of the land and sea . . . and the changes caused indirectly might ramify through all classes of the living creation, and be almost endless."

For Darwin, the tameness of the Galápagos birds became a symbol of Lyell's lesson. That is how Darwin closes his chapter on the Galápagos in his *Journal of Researches*, after describing the tameness of the birds. "We may infer from these facts," he writes, "what havoc the introduction of any new beast of prey must cause in a country, before the instincts of the aborigines become adapted to the stranger's craft or power."

Darwin elaborates the point in a famous passage in the *Origin*, a passage that helped establish the science of ecology. A "web of complex relations" binds all of the living things in any region, Darwin writes. Adding or subtracting even a single species causes waves of change that race through the web, "onwards in ever-increasing circles of complexity." The simple act of adding cats to an English village would reduce the number of field mice. Killing mice would benefit the bumblebees, whose nests and honeycombs the mice often devour. Increasing the number of bumblebees would benefit the heartsease and red clover, which are fertilized almost exclusively by bumblebees. So adding cats to the village could end by adding flowers.

For Darwin the whole of the Galápagos archipelago argues this fundamental lesson. The volcanoes are much more diverse in their biology than their geology. The contrast suggests that in the struggle for existence, species are shaped at least as much by the local flora and fauna as by the local soil and climate. Why else would the plants and animals differ radically among islands that have "the same geological nature, the same height, climate &c"?

"This long appeared to me a great difficulty," Darwin writes toward the end of the *Origin*, "but it arises in chief part from the deeply-seated error of considering the physical conditions of a country as the most important for its inhabitants; whereas it cannot, I think, be disputed that the nature of the other inhabitants, with which each has to compete, is at least as important, and generally a far more important element of success." When each seed and bird arrived on one of these islands, he says, "it would have to compete with different sets of organisms . . . and it would be exposed to the attacks of somewhat different enemies. If then it varied, natural selection would probably favour different varieties in the different islands."

Again, Darwin never actually saw this happen. But he concluded that the introduction of a new species anywhere on earth might be an evolutionary event: that small invasions can have far-reaching consequences. The invader itself evolves rapidly as it adapts to its new home; meanwhile, everything already living there either adapts to the invader or becomes extinct. The result is an expanding pulse of evolution and extinction, a general acceleration of the pace of change.

In the past, the pace of travel was generally slow. The finch had to wait for the freak wind. The polar bear had to wait for the iceberg. The union and disunion of continents were limited by the pace of continental drift, which is measured in centimeters per year.

Today invasions on this scale are happening every day, and virtually every spot on the surface of the planet is being assaulted by varied, new evolutionary pressures. Invaders ride in the puddles of old car tires, the bilge water of ships, the pressurized cabins of airplanes; in suitcases, on pant legs, and on mud-caked shoe soles. These are Darwin's old jelly-jar experiments replayed in earnest. The consequence is an intensification of Darwinian pressures not only in the Galápagos but everywhere we look.

THE YANKEE EVOLUTIONIST Hermon Carey Bumpus witnessed one small step of an invasion at the close of Darwin's century. On the last day in January 1898, Providence, Rhode Island, was struck by a howling blizzard. The storm blocked the street cars and the railroad trains, snapped the telegraph and telephone lines, and threw sparks from dangling wires onto the shingled roof of Walter H. Willis's furnishings store. (According to the front-page story in the next day's Providence *Journal*, the fire was put out with snowballs by "a crowd of merry makers.")

Bumpus was teaching biology at Brown University. His path to work each morning led up College Hill and past the Providence Athenaeum, one of the oldest libraries in the United States. The morning after the storm, as he made his way through the snow, he happened to notice a large number of English sparrows lying dead or exhausted in the drifts beneath the Athenaeum. The sparrows had been wintering in the ivy that covered the library, and the gale had overwhelmed them.

These sparrows were newcomers to New England, as Bumpus knew: Old World birds in the New World. One of the first pairs had been released in New York's Central Park in 1851, the decade before Bumpus was born, by an eccentric bird lover who wanted to import every one of the birds in Shakespeare's plays to the United States. So the birds were lying in the snow that morning in part because Shakespeare had written, "There is a special providence in the fall of a sparrow."

Bumpus collected as many of the sparrows as he could and brought them to the Anatomical Laboratory. In the warmth of the lab, seventy-two of the sparrows revived; sixty-four died. He recorded the sex, body length, wingspans, and weight of both the living and the dead; he also measured the length of the head, the humerus, the femur, the tibiotarsus, the skull, and the sternum. When he tabulated all these results he found that most of the survivors were males, and that among the males the survivors tended to be shorter and lighter than average, with "longer wing bones, longer legs, longer sternums, and greater brain capacity."

At the time, English sparrows were multiplying and overrunning the continent. Bumpus believed that "the English sparrow, since its introduction into this country, has found life so easy that the operation of natural selection has been practically suspended, and that the American type consequently has become degenerate." The birds that had died in the storm were those that had diverged most from the original type. It was an episode of what is now called stabilizing selection.

In the early 1970s, just before his first trip to the Galápagos, Peter Grant reread Bumpus's paper and studied the tables of numbers. The paper had long since become one of the most famous studies ever conducted of evolution in action. Peter reanalyzed the data using more powerful statistical tools. He concluded that Bumpus had actually seen not one but two kinds of natural selection. For the female sparrows the storm was stabilizing. The event killed the largest and the smallest but preserved the mean, just as Bumpus had said. In the males, however, the pressure of the storm was directional, pushing the birds toward smaller size.

The reanalysis of Bumpus's classic data helped to inspire the

Grants' first trip to the Galápagos. There was a providence in the fall
of those sparrows. If you could demonstrate natural selection under
those freakish conditions, Peter and Rosemary wondered what they
could see under more normal conditions. They might not be able to
see it at all.

Today English sparrows are still evolving and adapting to North
America (and to South America, South Africa, Hawaii, Australia, and
New Zealand) through Darwin's process. So are barnacles that invaded
the Salton Sea during World War II, riding in with the U.S. Navy. So
are fruit flies that invaded Chile in the 1970s. With these species and
many others, biologists are now tracking evolution in action, quite
confident—as the Grants' and other studies pile up—that the process
can be watched. Each of these creatures, now that we have put it
down in a new place in the world, is in the middle of a crisis, an ep-
ochal event, like the first finches to land in the Galápagos, and their
evolution, like the finches', is turning out to be rapid.

WHENEVER WE INTRODUCE AN ALIEN to a new country we
also change life for the natives—Lyell's point in the parable of the polar
bear, and Darwin's in his vision of cats helping flowers. This kind of
chain of events is going on everywhere too, and it too is being
watched.

The evolutionist Scott Carroll of the University of Utah is proving
the point with the beak of the soapberry bug. This is a bug with a
long needle of a proboscis—which is, in effect, its beak. The bug
pokes its beak through the walls of fruit, pierces the walls of the seeds,
and then liquefies and sucks up the nutrients, as if it were sipping cider
through a straw.

Carroll, who is one of the tallest biologists in the world, has spent
many of his working hours in the last few years stooping over soapberry bugs from Yavapai Country, Arizona, to Key Largo, Florida.
They are New World bugs, rather pretty-looking things. Carroll paints
black numbers on their backs to keep track of individuals and watches
them eat. At the start of his study, he says, he assumed they would be
like little wind-up automata, but they are not: privately he has come
to believe as he watches them that they are each different, with per-
sonalities. But that is off the subject of his quest.

In the south-central United States, soapberry bugs live on the
soapberry tree. In southernmost Texas they live on the *serjania* vine. In

southern Florida they live on the perennial balloon vine. These are its native hosts; the bug and these plants go back thousands of years.

Besides these three native plants, the bugs also steal seeds from three species that have been introduced to their turf quite recently: the "round-podded" golden rain tree and the "flat-podded" golden rain tree, which have been imported from Southeast Asia as ornamentals, and the heartseed vine, which grows wild in Louisiana and Mississippi.

The beak of the soapberry bug is easy to measure, and the meaning of the variations is obvious. Someone once asked Abe Lincoln how long he thought a man's legs should be. Abe is said to have replied, "Just long enough to reach the ground." Likewise the beak of the soapberry bug should be just long enough to reach the seeds.

Now, the fruit of the balloon vine, which is native to South Florida, has a radius of a dozen millimeters. The beak length of the soapberry bugs that feed on it is a little more than 9 millimeters, long enough to reach the seeds, which cluster around the center of the fruit. But the fruit of the flat-podded golden rain tree, which is new to Florida, has a radius of less than 3 millimeters. The soapberry bugs that eat these fruits do not need such long beaks, and accordingly their beaks are growing shorter: they now average less than 7 millimeters long. That is rapid evolution, since the golden rain tree was not planted in any significant numbers in Florida until the 1950s (several thousand generations, for the bugs). With time their beaks are likely to grow shorter yet.

Meanwhile the fruit of the native soapberry tree has a radius of only 6 millimeters, and the beaks of the bugs that dine on it are 6 millimeters long. But the fruit of the newly introduced heartseed vine is almost 9 millimeters. Soapberry bugs that have switched to heartseed now have beaks almost 8 millimeters long. That is even faster evolution, for heartseed did not become abundant in the territory of soapberry bugs until about 1970. With time their beaks will grow even longer.

Carroll has also measured specimens of these bugs in state and county natural-history museums. In case after case, he finds a change in beak length at about the right time—that is, about the time the new plants were introduced in their territory.

In every case the beaks of the soapberry bugs are evolving to obey Lincoln's rule: just long enough to reach the seeds. And because the dates of introductions of the new plants are known so well, and are so recent, it is clear that this is yet another example of evolution in action around us, with the driver and the driven both plainly in view. Carroll calls this "natural history with the history."

DARWIN SUSPECTED that this kind of change in habits might sometimes lead in surprising directions. In the *Origin* he notes that there are species of British insects that feed exclusively on British farmers' crops. Meanwhile, the insects' ancestral line goes on eating the native weeds in the woods and hedgerows around the farmers' fields. Darwin concludes that "within the same area, varieties of the same animal can long remain distinct, from haunting different stations, from breeding at slightly different seasons, or from varieties of the same kind preferring to pair together." He thought that many new and "perfectly defined" species might have been formed this way, not on islands but side by side in the fields and hedgerows of rural England.

One of the *Origin*'s first readers in the New World was a British graduate of Cambridge from the same year as Darwin. Benjamin Walsh had studied for the church in Cambridge, and like Darwin he had discovered that he preferred bug collecting. Instead of setting up in a country estate outside London, as Charles and Emma did, Walsh and his wife sailed for the United States.

In Illinois Walsh "built a mud-plastered log cabin with a fireplace that took a log that had to be rolled in with oxen," as he later wrote to Darwin at Down:

> I was possessed with an absurd notion that I would live a perfectly natural life, independent of the whole world—*in me ipso totus teres atque rotundus.* So I bought several hundred acres of wild land in the wilderness, twenty miles from any settlement that you would call even a village, and with only a single neighbor. There I gradually opened a farm, working myself like a horse. . . .

By the time Darwin published the *Origin*, Walsh had caught malaria, lost two farms, built a lumber business, built ten brick tenements, stormed into and out of local politics, and settled back at last into entomology. In 1864 he sent a letter to England:

> More than thirty years ago I was introduced to you at your rooms in Christ's College . . . and had the pleasure of seeing your noble collection of British Coleoptera. Allow me to take this opportunity

for thanking you for the publication of your *Origin of Species.* . . . The first perusal staggered me, the second convinced me, and the oftener I read it the more convinced I am of the general soundness of your theory.

Walsh applied Darwin's ideas to what he saw happening in the fields and woods around him. There is a species of native American fruit fly, commonly known as the haw fly, that lays its eggs on wild hawthorn. In Walsh's day, farmers had begun planting and cultivating apple orchards in the Hudson River Valley. Some haw flies had left the haws and begun eating the apples.

Although this species of fly "exists both in the East and in the West," Walsh wrote in 1867, "it attacks the cultivated apples only in a certain limited region, even in the East, for . . . this new and formidable enemy of the apple is found in the Hudson River Valley, but has not yet reached New Jersey." He predicted that the flies would continue to spread, and that they might eventually begin to diverge. The haw flies and the apple flies might go on living side by side, yet turn into two separate species, as Darwin suggests in the *Origin.*

Unfortunately, Walsh died soon after publishing his paper on the apple fly; but the flies did spread as he predicted. They were reported north of the Hudson River Valley, in Vermont and New Hampshire, in 1872, in Maine in 1876, in Canada in 1907. By 1894 they were marching through Georgia, and by 1902 they were in Michigan. They are now eating apples up and down the East Coast of North America and in the Midwest; in the last decade they arrived on the West Coast. The fly also eats rose hips, and sometimes pears and plums, and in the Door Peninsula in Wisconsin it has just begun eating sour cherries. Meanwhile haw flies still eat haws.

Currently most evolutionists regard the possibility of speciation among neighbors as unorthodox, even though Darwin himself proposed it. The standard model of speciation requires geographic isolation. That has been the canonical pattern for half a century, and many evolutionists believe it is the universal pattern. But evolutionists are forever dividing and subdividing into schismatic sects, kingdoms of Either and Or. Do new species arise in archipelagoes, like Darwin's finches, or do they arise among neighbors? Is the origin of species fast or slow? Is the mechanism natural selection or sexual selection? And so on. None of these questions really have to be framed either-or. It is al-

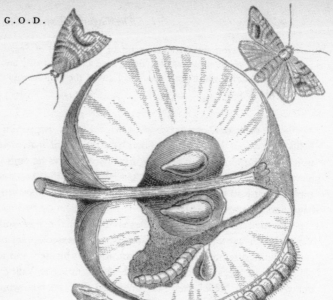

most a law of science: the more indirect the evidence, the more polarized the debate. Evolutionists sometimes catch themselves sounding like the Little-Endians and Big-Endians in *Gulliver's Travels*, fighting tooth and nail over the proper way to crack an egg. Meanwhile, the more direct the evidence, the less the answers look either-or.

Flies that have switched from haws to apples lay their eggs in an apple while it is still on the tree. The eggs hatch within two days, and the larva grows and grows, hollowing and honeycombing the fruit as it eats. When the apple ripens and falls to the ground, the larvae dig down into the soil. They spend the winter dormant in the ground at the foot of the tree they infested the summer before. By the time the next summer comes, they are ready for another crop of apples.

The flies mate only on the trees where they feed. Early in the season, according to one apple-fly expert, the male flies toward the female and stares at her, face to face, within one or two centimeters; "and then, if the gestalt of visual characteristics suggests that the insect is indeed of the same species, jumps onto [her] abdomen and attempts to copulate." Often, on top of an apple.

Later in the season, the courtship pattern changes. The males establish territories and defend their turf from other males while they wait for females to alight nearby. Two apple flies sometimes battle each

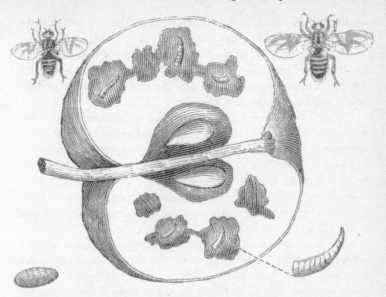

The apple on the left has been attacked by a codling moth, the apple on the right by an apple fly. From Benjamin D. Walsh, "The Apple-Worm and the Apple-Maggot." *Library, the Academy of Natural Sciences of Philadelphia*

other for possession of an apple. So these flies' lives seem to be growing more and more entwined with the apple trees. They may be in the act of cutting themselves off from the haw flies and forming a new species, just as Walsh predicted a century ago.

A young evolutionist, Jeffrey Feder, is now investigating this possibility at the molecular level. In one study he looked at haw and apple flies in an old, abandoned orchard near the town of Grant, Michigan. He and his coworkers examined six enzymes in the flies. For all six enzymes, Feder found slight differences between the flies on the haws and the flies on the apples. The differences really are extremely slight. In fact, Feder found precisely the same six enzymes in both the haw flies and the apple flies. He also found the same variant forms of the six enzymes. The haw and apple flies differed only in the *proportions*, the relative numbers, of the variant forms. It is as if human beings on two continents each carried the same set of genes for eye color—blue, brown, black, green—but on one continent blue was more common than black, and on the other black was more common than blue. The gene pools of the two races of flies in the orchard seem to be partly joined and partly isolated. We may be watching the very earliest stages in the divergence of two species.

In a more extensive study, Feder looked at two dozen enzymes in flies from a dozen orchards across the United States and Canada. He saw that all over the continent the gene pools of the apple race and the haw race have diverged in this way.

Apples ripen and fall about one month earlier than haws. When the flies tie themselves to the apples, their feeding and breeding schedules have to shift about a month ahead in the calendar. One month is a lifetime for these flies: it is nearly the entire life-span of an adult fruit fly. So the switch to apples is pushing the two gene pools apart. Haw flies and apple flies are marooned on islands in time rather than islands in space. They seem to be taking the first steps toward speciation, just as Darwin and Walsh imagined.

The same thing must be happening unseen in thousands of other species around the world, because farms offer new islands and niches for insects everywhere, as every farmer knows and regrets. Codling moths too have adapted to North American apple orchards since the moths arrived in the Hudson River Valley two hundred years ago. Other populations of codling moths specialize in Persian walnuts, Swiss apricots, South African pears, California plums. Treehoppers have diverged and branched in six divergent directions, to specialize in six different genera of trees and bushes, including bittersweet; butternut, *Viburnum*, and *Cercis*. Treehoppers in each genus of tree do not mate with treehoppers in another tree, even if the two trees grow side by side. The reproductive isolation of the treehoppers is complete.

Feder has also studied a line of flies that looks identical to haw flies and apple flies but infests blueberries and huckleberries. It is a menace in blueberry patches from Nova Scotia to Florida, and as far west as Michigan. If you take some apple flies and some blueberry flies and put them in a jar together, they will mate and produce perfectly normal-looking, healthy, hybrid flies. They appear to be absolutely interfertile in the jar. But Feder has inspected the genes of adult flies collected from neighboring blueberry bushes and apple trees. He finds that their genes are clearly distinct and "species-specific," peculiar to the apple fly or the blueberry fly. So they hybridize rarely, if at all, in the wild. They are not passing genes back and forth between them. Even where the highest branches of a blueberry bush and the lowest branches of an apple tree are interdigitated, mingling like the fingers of two hands, the two species of flies are still not mixing. Each fly is feeding on the fruit of its kind, and mating true to its kind, as isolated as if the blueberry and the apple flies lived on far-flung islands.

Chapter 17

The Stranger's Power

We may infer from these facts, what havoc the introduction
of any new beast of prey must cause in a country, before the
instincts of the aborigines become adapted to the stranger's
craft or power.

— CHARLES DARWIN,
Journal of Researches

In the Galápagos, large tracts of the highlands of Floreana, Santiago,
San Cristóbal, and Santa Cruz have been cleared for cattle pas-
ture. Farmers have planted tomatoes, avocados, guavas, papaya, oranges, lemons, bananas, potatoes, cabbages, coffee, and a fruit known
locally as Norwegian pear. On Santa Cruz, which is in sight of
Daphne Major, the dirt road up through the highlands and over the
summit of the island is full of wide views of these young orchards and
pastures, and dotted with small white crosses, like wayside shrines, to
commemorate fatal car accidents. Horses, donkeys, cows, and cattle
egrets (all of them, like human beings, new to the islands) stand in the
fields, through which a few giant Galápagos tortoises still wander or
browse in green pond scum with Darwin's finches riding on their
domes. The tortoises help the farmers by eating grapefruit, oranges,
and lemons whenever they can, and dispersing the seeds throughout
the uplands.

The biggest settlement in the archipelago is the village of Puerto
Ayora, on the southern shore of Santa Cruz. Darwin's finches are the
sparrows of the village. In the cafés, while men call greetings to one
another, mock-distinguished epithets—*"Capitán!" "Profesor!"*—the
finches hop beneath their feet, looking for crumbs and spills. They
hunt and peck in every dooryard garden and bathe in puddles in the
middle of the dirt roads, braving the bicycles and pick-up trucks.

The local people call the finches *chiques*, for the sound. They know the birds are famous, but they don't know why. They simply call them *chiques*—all of the finches, from the small-beaked to the big-beaked; the warblers are *marias* or *canarios*.

Darwin knew that the finches could adjust their habits quickly to take advantage of the banquets spread by human beings. In the highlands of Floreana he watched the birds stealing seeds from the fields of the Ecuadoran political prisoners. "The Gross-beaks are very injurious to the cultivated land," Darwin writes in his *Ornithological Notes*; "they stock up seeds & plants, buried six inches beneath the surface."

Similarly, Alfred Russel Wallace in an essay on the variability of species tells the story of the kea, a mountain parrot of New Zealand. Before Europeans began raising sheep in New Zealand, the kea fed on the honey of flowers, and on the insects that buzzed around the flowers, rounding out its diet with fruits and berries. Soon after shepherds came to New Zealand, keas began picking at sheepskins hung out to dry and the mutton hung up to cure. Then shepherds began finding sheep with raw and bleeding wounds on their backs. About 1868, Wallace says, some of the shepherds actually saw the parrots attacking their sheep. "Since then," Wallace reports, "it is stated that the bird actually burrows into the living sheep, eating its way down to the kidneys, which form its special delicacy." The shepherds meanwhile declared war on the parrots, shooting them on sight. "The case," Wallace concludes, "affords a remarkable instance of how the climbing feet and powerful hooked beak developed for one set of purposes can be applied to another altogether different purpose, and it also shows how little real stability there may be in what appear to us the most fixed habits of life."

In the Galápagos the lives of the settlers and Darwin's finches have almost instantly intertwined. "Finches eat the flowers of the crops," says Fabio Peñafiel, a young Ecuadoran naturalist and Galápagos guide who lives in the village of Puerto Ayora. "The *señora* who lives next door to me has had to give up planting pineapple in her garden," he says, "because of the finches picking in the middle. Cactus finches and tree finches. It's nibble, nibble, nibble, and no fruit will come out.

"The *señora* has eaten finches. Yes! She says they have a very mild taste. She makes finch soup. And of course she tells me that! She says, 'Your finches, your silly finches have eaten my plants—but I have eaten the finches!'

"Highly illegal! Finch soup! The people at the station would just fall on their backs! But she was honest enough to tell me. Imagine what the people up in the highlands are doing, when the finches are eating their crops. And after all, the finch population is probably increasing up there because of those crops. I don't like sentimentality. Facts are facts."

At the Charles Darwin Research Station, which is just outside the village, Darwin's finches hop in the gravel between the low buildings, hunting for seeds among the pebbles and the pocked pale coral. They look down at visiting scientists from the edges of the corrugated roofs and the railings of the dormitory porches. They perch on the arrows along the tourist footpaths as if seeking evolutionary guidance. The flocks of finches are thickest outside the station's library, where the li-

Darwin's finches on a tourists' footpath at the Charles Darwin Research Station, in the village of Puerto Ayora.
Drawing by Thalia Grant

brarian, Gayle Davies, throws rice to them, wearing a shirt tied at the waist and hair pulled back in a bun, tropical-librarian-fashion. She also feeds the finches on the veranda of her home, a mile down the dirt road from the station. There, she dumps the rice on a hanging tray, while dozens of *fuliginosa, fortis, magnirostris,* and *scandens* hop and cling to the wires, cheeping and snatching a few grains right out of her hands.

"Since my husband is gone a lot—he goes out to sea very often—it's nice to have living things around," Gayle says. "They are there in the morning. They're there at noon. And they are *hopeful* all day long. If someone comes to the house when I'm away, it's like spreading a rumor: all of a sudden there's a flock."

She used to keep the rice in an open bowl in the kitchen, and some of the finches, the most confident, would fly right in and feed from the bowl. Most preferred to wait until they were served outside. "But some few who didn't like to eat with the hoi polloi would come right through the window to the inside bowl and eat in peace, even if the outside bowl was full," she says.

"This one day, I was inside the house, sitting on the bed reading. The house we lived in then was just one room—the bed doubled as a couch by day. So I'm sitting propped up on pillows, and a finch landed on a pillow right next to my head. And I could see that he had something wrong with his beak. It was pox. Pox shows normally on feet, but sometimes you get a nasty growth on the beak, inside.

"I managed to scrape it off. Then I painted the spot with gentian violet. I tried to help him out.

"I never had a finch do that before—fly right up and look into my face. They cock their heads and look at you when you are feeding them, but this was something very different. It felt—to get thoroughly anthropomorphic—like a tiny cry for help. Of course you'll never know. It could have been that he could not eat, was basically starving, and I was the source—the one who put out the rice. Who knows? To me, it felt like a 'help me.' "

THE ARRIVAL OF HUMAN BEINGS means a new phase in the evolution of Darwin's finches, and its directions are still unclear. "On the inhabited islands," says Peter Grant, "we haven't done studies of the effects of cats, rats, mice, dogs, goats, donkeys, fire ants, pineapples, bananas, or guavas—and so on. Are they having an effect on the finches? Strictly speaking, we don't know from our studies. We assume they are," he says, "but we have *observed* none of these changes."

Rosemary and Peter do think they see something odd about the finches of Santa Cruz. The birds around the research station, and in the village, seem to be blurring together. The Grants have never made a systematic study of this: but to their eyes the species almost look as though they are fusing. "They just sort of run into each other," says Rosemary. There is no difference between the largest *fortis* and the smallest *magnirostris*.

"Really, we would need to compare the finches one hundred years ago in that part of the island with what they are now," Peter says cautiously. But he and Rosemary wonder if the provision of so much water and food in Puerto Ayora may have enabled the birds' breeding densities to rise around the village. The struggle for existence may have grown less intense. As the growth of the village relaxes the selection pressures on generation after generation of finches, the birds may be turning into a hybrid swarm.

This has not happened on Daphne or Genovesa because, as the Finch Unit has shown, the struggle for existence on those uninhabited islands is intermittently so intense. In hard times the varied finch beaks have such great adaptive value that they are preserved again and again by natural selection. On Genovesa, for instance, late in 1983, after the great Niño, the island was unusually lush and conditions were unusually soft for Darwin's finches, as they are all year round on Gayle Davies's veranda. At that time the depths of *conirostris* and *magnirostris* beaks on Genovesa were very variable, and the deepest *conirostris* beak and the shallowest *magnirostris* beak were close together.

If conditions had continued lush, those finches would likely have continued to fuse. But the years 1984 and 1985 were one long dry period. "All the intermediates went out," says Rosemary. "Died, to be more precise: all the finches whose measurements lay between *conirostris* and *magnirostris* died."

In *magnirostris*, length and depth went up. But in *conirostris* they went down. As a result, the two species were divided by the selection episode, says Rosemary, gesturing with her hands as if she were drawing out a lump of clay. "The net result was divergence. You got an actual gap."

So the Grants speculate that in places like Santa Cruz, where there is little selection against hybrids, some of Darwin's finches might yet merge. Human beings may have tipped the balance between fission and fusion. Around the villages and farms, many of the celebrated differences in their beaks may simply disappear.

SOME DECADES AGO, the botanist Edgar Anderson wrote a speculative essay, "The Hybridization of the Habitat." Anderson is the evolutionist who coined the term introgressive hybridization, to emphasize the point that the backcrossing of hybrids with the lines of either of their parents provides a means for the mixing of genes between the two lines, and so may be an important evolutionary step.

In his essay, Anderson argues that the disturbances that human beings are visiting on the planet must be leading to increasing cases of hybridization everywhere, and that thousands of these hybrids and their habitats may prove to be the seedbeds of new evolutionary lines. In his view the crucial evolutionary step comes not with the first hybrid generation, but with the second. "The first hybrid will be uniform in its requirements and on the whole they will be for conditions intermediate between those required for the two parents," Anderson writes. This is just what the Grants are seeing with Darwin's finches: the first generation of hybrids is intermediate in beak and body. But from their offspring, surprises may come. "The second generation will be made up of individuals each of which will require its own peculiar habitat," Anderson writes, and repeats the statement in block letters: "THE SECOND GENERATION WILL BE MADE UP OF INDIVIDUALS EACH OF WHICH WILL REQUIRE ITS OWN PECULIAR HABITAT FOR OPTIMUM DEVELOPMENT."

In other words, the offspring of hybrids will be novel and bizarre, and they will require habitats as novel and bizarre as they are themselves. Anderson showed how this might happen in a study of two species of spiderwort that grow wild in the Ozarks. One grows in deep dark woods at the foot of cliffs, he says, while the other grows in full sun at the top of the cliffs. In many ways these habitats are polar opposites. The species at the base of cliffs requires:

rich loam
deep shade
leaf-mold cover

while the species at the top of the cliff requires:

rocky soil
full sun
no leaf-mold cover

The two spiderworts, he says, "are well-differentiated species; neither one of them is by any means the closest relative of the other." He could cross them readily in experimental gardens, where he learned to recognize the distinctive appearance of the hybrids. But he seldom found one of the hybrids growing in the wild, because the Ozarks rarely provided an intermediate habitat (not even the very middle of the cliff would do).

Anderson did find a few of these hybrids growing wild in the woods. Imagine what would happen if these hybrids crossed, he writes. Their offspring, by shuffling their diverse sets of genes, would now require six new habitats, besides their parents' two:

rich loam	rocky soil
full sun	deep shade
no leaf mold	leaf mold

rich loam	rocky soil
full sun	full sun
leaf mold	leaf mold

rich loam	rocky soil
deep shade	deep shade
no leaf mold	no leaf mold

The actual number of differences between any two species is much greater than in this schematic example. What is more, Anderson points out, the number of different habitats the hybrids require "will rise exponentially with the number of basic differences between the species. With ten such differences, around a thousand different kinds of habitat would be needed to permit the various recombinations to find a niche somewhere. . . . With only twenty such basic differences (and this seems like a conservative figure) over a million different recombined habitats would be needed."

You don't find situations that chaotic under natural conditions, but you do find them in the havoc that human beings bring in their train. Our arrival, Anderson says, "can provide strange new niches of hybrid recombinations." Thus, our disturbances hybridize both the environment and the species. We are hybridizing the planet.

Botanists have actually watched this happening. Farmers on the Mississippi Delta treat their lands slightly differently, and botanists have discovered that different wildflower hybrids flourish in the different conditions on each farm. The land is the same, the climate is the same, but the hybrids, as Anderson writes, sometimes go "right up to the fenceline at the border of the farm and stop there." The same sort of pattern has been documented in the fields and pastures of the St. Lawrence Valley and elsewhere. One botanist has looked at two species of sage that grew in the chaparral of the San Gabriel Mountains. In the chaparral itself he found no hybrids. But he did find them flourishing next to it, in an abandoned olive orchard. "In this greatly disturbed area, new niches were created for the hybrid progeny, which are apparently always being produced in the chaparral but at very low frequency," Anderson writes. "In this strange new set of various habitats some of the mongrels were at a greater selective advantage." The sage in the deserted olive orchard was composed of hybrids, and almost none of the original species grew there.

This kind of thing must have happened whenever new islands were colonized by windblown or seabird-carried seeds, as in the Galápagos. It must have happened whenever new continents were colonized by new plants and animals. "At such times and places," Anderson argues, "introgressive hybridization must have played an important role in evolution."

In a paper he wrote with G. Ledyard Stebbins, "Hybridization as an Evolutionary Stimulus," Anderson argues that what we are seeing now in the world is something the world has seen in one form or another many times before. We are ecological dominants, the two evolutionists write, but we are not the first to bestride the world in conquest. "When the first land vertebrates invaded terrestrial vegetation they must have been quite as catastrophic to the flora which had been evolved in the absence of such creatures. When the large herbivorous reptiles first appeared, and also when the first large land mammals arrived in each new portion of the world there must have been violent readjustments and the creation of new ecological niches."

This would have happened in the Hawaiian Islands, the Galápagos, Lake Baikal, wherever "species belonging to different faunas and floras were brought together" and "physical and biological barrier systems were broken down." It would have happened most recently in the upheavals of the Pleistocene when ice sheets swept across the continents of the Northern Hemisphere and there were icecaps even on the summits of the island of Hawaii.

So in chaotic times, after each deluge, hybridization may help the standard Darwinian mechanisms to run faster. By intensifying the pressure of selection everywhere on the world and disturbing habitats all over the world we may be creating conditions in which evolution is running at its maximum rate. Anderson and Stebbins argue that "the recent rapid evolution of weeds and semi-weeds is an indication of what must have happened again and again in geological history whenever any species or group of species became so ecologically dominant as greatly to upset the habitats of their own times."

Under our influence, as Anderson and Stebbins write, "evolution has been greatly accelerated. There has been a rapid evolution of plants and animals under domestication and an almost equally rapid evolution of weed species and strains in greatly disturbed habitats."

Normally we think of the intrusions of weeds and introduced species as distractions from the great work of nature, the creative process of Darwinism. But this kind of upheaval has happened again and again, though never through the advent of a conscious dominant like ourselves, a conqueror come to watch. So what we are seeing is a specimen of evolution as it has happened again and again throughout the ages. "Far from being without bearing on general theories of evolution," write Anderson and Stebbins, these events "are of tremendous significance, showing how much more rapidly evolution can proceed under the impact of a new ecological dominant (in this case, Man)."

"The enhanced evolution which we see in our own gardens, dooryards, dumps and roadsides may well be typical of what happened during the rise of previous ecological dominants," they write. We are doing what the dinosaurs did before us, only faster. We bring strangers together to make strange bedfellows, and we remake the beds they lie in, all at once.

DARWIN AND HIS shipmates collected fifteen finches on the island of Floreana. Five of them were an unusually large race, *magnirostris magnirostris*: the biggest of the biggest of Darwin's finches. They must have been extremely common on the island at the time; in old owl lairs on Floreana, the bones of this giant among finches are more than twelve times more abundant than any other finch.

The evolutionist David Steadman has made a study of the great

Galápagos finch. He notes that when Adolphe-Simon Neboux and Charles-Rene-Augustin Leclancher of the French frigate *Venus* collected birds on Floreana in 1838, only three years after Darwin, they did not find a single *magnirostris magnirostris*. Neither did Thomas Edmonston of HMS *Herald* in 1846, or one Dr. Kinberg of the Swedish frigate *Eugenie* in 1852. In fact *magnirostris magnirostris* has never been seen since. It seems to have gone extinct almost immediately after Darwin's visit to the islands. Darwin was the first naturalist to collect a specimen and the last to see one alive.

What happened to the giant finch of Floreana? No one on the island was observing it the way the Finch Unit is now watching Daphne. But it is certain that the birds were not wiped out by a volcanic eruption, because Floreana has been dormant for as long as human beings have been in the islands. The finches almost certainly died because of another kind of eruption, the upheaval that human beings brought to their island. The culprit in their extinction is almost certainly the prison colony, which had been established only a few years before Darwin's arrival. There were two or three hundred Ecuadoran prisoners on the island when the *Beagle* stopped there, as Darwin recorded in his journal. They had cleared farms in the highlands where they tended sweet potatoes and plantains. They hunted wild pigs and goats and feasted on the giant tortoises, and at least one family ate finch soup.

In some ways the arrival of the farmers was a bonanza for the finches. They made pests of themselves on the farms, as they are doing now on Santa Cruz. The "Gross-beaks" that Darwin saw digging up seeds in the fields were almost certainly *magnirostris magnirostris*.

Unfortunately for the finches, however, the prisoners imported more than seeds to the island. They also brought cattle, goats, pigs, cats, and rats. And though the prisoners were soon gone (the colony lasted only a few years after Darwin's visit), their retinue of animals remained and multiplied. When Captain A. H. Markham of HMS *Triumph* visited the islands in 1880, he found it to be "in undisturbed possession of the so-called wild cattle." He also saw "donkeys, dogs, pigs, and other animals that had been left to run wild on the abandonment of the island by the former inhabitants."

Magnirostris magnirostris were big birds, and probably didn't fly much, even by the standards of ground finches, says Peter Grant. They must have been easy for rats and cats to jump on. To make matters worse for the finches, the local cactus went extinct. *Magnirostris*

magnirostris was probably dependent on this cactus, which was an equally unusual local race called *megasperma megasperma*, meaning big, big seed. The seeds of this cactus are the largest and hardest of any cactus in the Galápagos, about half an inch in diameter. A relic population of the giant cactus survives on a small outlying island, Champion, off the coast of Floreana. The Grants have tested its seeds with their McGill nutcracker. It rates a 20 on their scale for size and hardness, while the next-toughest cactus seed in the Galápagos rates only just under 11. *Megasperma megasperma* produces these giant seeds in prolific numbers. Giant *mags* probably could crack those seeds, says Peter Grant, and for a relatively small investment of time, get a big return. But any ordinary finch would find them impossible to crack.

Almost certainly the doomed *magnirostris magnirostris* specialized in *megasperma megasperma*. In fact their beaks and the cactus seeds may have evolved in a kind of arms race, the finches' predations driving the cactus to make bigger and bigger seeds, and the bigger seeds driving the finches to bigger and bigger beaks.

When the prisoners left Floreana their abandoned cattle and donkeys learned to knock over the cactus and chew its flesh for water. In a very short time there was almost no *megasperma megasperma* left on the island. Without it, conditions for *magnirostris magnirostris* may have become unlivable. Mockingbirds went extinct on Floreana too, at about the same time. Galápagos mockingbirds nest in cactus.

TODAY THE SCIENTISTS at the research station are trying hard to rescue many of the best-beloved animals in the Galápagos, including iguanas, tortoises, and dark-rumped petrels. Living things on islands are more vulnerable than their relatives on continents, because they are small in numbers, and because they evolved apart from the dizzying welter of competition on the continents. Around the world almost one hundred species of birds are known to have gone extinct since the seventeenth century, together with more than eighty subspecies; more than nine out of ten of these lost species and subspecies lived on islands.

"If goats got onto Daphne, for instance, they would eat out the cacti, which would certainly drive *scandens* extinct," says Rosemary. "And if cactus and *Tribulus* were both wiped out, that would finish all the finches with big beaks and medium-sized beaks."

Cactus finch.
Drawing by Thalia Grant

That is why the Grants take such extraordinary precautions each time they travel from Santa Cruz to the Landing on Daphne, or to Darwin Bay on Genovesa. They do not want to bring even one pregnant fire ant with them.

"We'd be horribly upset—" says Peter.

"—Oh, my God, yes—" says Rosemary.

"—because that is not the island's natural state," says Peter. "And if we were responsible, we'd be devastated." Every time they land on the island, they wash all the fresh food in salt water, and the mist nets' poles too. Sometimes they tow the poles in the water all the way to the island, to make sure stowaway ants are drowned. "We always wash the tents and the bird bags, and we just sort of sanitize everything," says Rosemary. "Usually I take a can of spray just in case, but we've never had to use it."

"Fire ants would kill off all the scorpions, and probably the spiders," Peter says. "And since these are small islands, the scorpions and spiders would probably go extinct."

A lot of people would be happy to lose the scorpions and spiders. But then, as Peter retorts, "A lot of people live in cities."

"If they killed the scorpions, that would affect the *Tropidurus*, the lava lizards," Rosemary says.

"But it is the general principle, more than concern for any one species," says Peter. "If any new species invades, the old balance is gone."

"The changes would reach through the food chains," says Rosemary.

"Yes, changes of unpredictable sorts. It would be very difficult to be sure of the consequences for organisms higher up in the food webs. The habitats we are watching are so sensitive and fragile," Peter says. "It's easy to imagine major changes taking place on both islands."

"So few places have not been touched by humans," says Rosemary. "You can imagine if a big hotel, a Holiday Inn were built on Daphne or Genovesa—with the introduced plants, and insects, and rats, and cats. . . ."

"Parrots, budgerigars. . . ."

"You'd get Daphne and Genovesa just totally destroyed."

ON THE HAWAIIAN Islands, which were once the most isolated archipelago in the world, human beings arrived near the beginning of the first millennium. Many other aggressive species landed with us, in accelerating waves, including pigs, goats, rats, mongooses, mosquitoes, army worms, bookworms, cockroaches, centipedes, and scorpions, not to mention the English sparrows and more than 4,500 species of alien plants. Beneath these tidal waves the adaptive radiations of native birds in the archipelago have been crushed. Species of finches, hawks, owls, flightless ibis, and flightless rails have gone extinct; the finches have been hit particularly hard. "Even casual observers have noted a kaleidoscope of shifting dominance by different species of alien birds over the past thirty years," says one ecologist, Peter Vitousek; "the one constant has been the near absence of natives."

In the far north and west of the Hawaiian archipelago there is an atoll called Laysan. The adaptive landscapes of this remote island were turned upside down in the nineteenth century by the arrival of European rabbits and rats. In 1967, in a sort of Noah's Ark project, the U.S.

Fish and Wildlife Service took away more than one hundred specimens of an endangered species of Hawaiian finches and carried the birds by boat about 500 kilometers (more than 300 miles) to a small cluster of Hawaiian Islands known as Pearl and Hermes Reef.

After the Laysan finch was introduced to this reef, Sheila Conant, an ornithologist at the University of Hawaii, inspired by the Grants' work in the Galápagos, began watching the birds in their new home islands. The Laysan finches are as tame as Darwin's, and Conant and a few field assistants have been able to band most of them. She could not watch the birds almost continuously the way the Grants and their teams of watchers have managed to do. But she has made a few expeditions to the islands, spaced several years apart.

Conant found that within twenty years of their introduction to the islands of Pearl and Hermes Reef, the finches had begun to go off in different directions. On Laysan their ancestors had had shallow, wide,

A Galápagos finch hunts flies on an iguana.
Drawing by Thalia Grant

medium-length beaks. Now, in their new home, the finches on Southeast Island have longer beaks, while the finches on North Island have shorter, deeper, narrower beaks.

These are grain-eating finches. The seeds they take range from very small and soft to big and hard, the toughest of all being, once again, *Tribulus*, which probably got to Hawaii the same way it got to the Galápagos, hitching rides on sailors and whalers. *Tribulus* is not common on Laysan, but it is one of the main plants on Pearl and Hermes Reef. Correspondingly, the finches in their new home spend much more time working on *Tribulus*. And the *Tribulus* mericarps on Southeast Island are larger than on North Island, which may explain the divergence in the beaks of these finches.

The speed with which these finches have adapted is amazing, and it shows how fast the primoprogenitors of Darwin's finches may have adapted when they first came to the Galápagos. "Are we tinkering with evolution?" asks Sheila Conant. The answer is, of course, yes. As more and more species are endangered, more and more well-meaning attempts to save them involve introducing them to new homes or reintroducing them to their old homes. What happens next in these cases will seldom be monitored as closely as the Grants are watching Daphne, and Conant is watching Laysan. But it is clear from these studies that the introduced species, if it survives, can evolve rapidly and unpredictably in its new homes. To Conant this is intriguing and perplexing. Conservationists may be frustrated to find that by moving a species, instead of helping it to survive, they have helped it evolve into a new form.

That is why Rosemary and Peter put the word *dynamic* in the title of the first book they wrote together: *Evolutionary Dynamics of a Natural Population*. "It's important to keep in mind," says Rosemary: "Species don't stand still. You can't 'preserve' a species." Every species is, as the Grants write in the last words of their book, "constantly changing and capable of further change."

All of which brings home a "bitter-sweet message," as Peter Boag wrote in *Nature* on the centenary of Darwin's death: "much less is known about evolution in the Galápagos than most people think, but Galápagos populations and communities are probably now changing faster than ever."

Chapter 18

The Resistance Movement

Let a man profess to have discovered some new Patent Powder Pimperlimplimp, a single pinch of which being thrown into each corner of a field will kill every bug throughout its whole extent, and people will listen to him with attention and respect. But tell them of any simple common-sense plan, based upon correct scientific principles, to check and keep within reasonable bounds the insect foes of the farmer, and they will laugh you to scorn.

— BENJAMIN WALSH,
The Practical Entomologist (1866)

All summer long, messengers from Federal Express deliver large white parcels to a laboratory down on Moffett C-Level. The parcels come from hamlets in Louisiana, Southern California, and many states in between: the swath of North America that is often called the Cotton Belt, or the Bible Belt.

As these parcels come in, a postdoc in the laboratory sets them on his lab bench, slits the FedEx seals, and lifts the insulated lids. Inside, each box is molded in the shape of a white crater. In the bottom of the crater, lying on a bed of dry ice, amid a faint rising cloud of vapor, are a dozen gray moths.

The postdoc, Martin Taylor, lifts out the moths one by one with a tweezers. One by one he grinds them in a mortar and pestle, like an old-fashioned pharmacist, to extract their DNA. He is watching these moths from his laboratory bench the way Peter Boag is watching Darwin's finches. Some of the greatest opposition to evolution comes from the farmers of the Cotton Belt, and that is where Taylor is seeing one of the most dramatic cases of evolution in action on this planet.

There was a time, not so long ago, when this particular species of moth, *Heliothis virescens*, lived a life of obscurity, probably in forests and hedgerows, where it ate weeds. But in the year 1940, cotton farmers began spraying their fields with the chemical compound dichloro-diphenyltrichloroethane, better known as DDT. These first insecticidal sprays killed so many insects, and killed so many of the birds that ate the insects, that in biological terms the cotton fields were left standing virtually vacant, like an archipelago of newborn islands—and out of the woods and hedgerows fluttered *Heliothis virescens*.

Some of these moths were able to tolerate DDT. They lived long enough to lay their eggs in the cotton bolls. Their eggs hatched into larvae, and the larvae began devouring the cotton.

In the next few optimistic years, pesticide manufacturers assaulted *Heliothis* with bigger and bigger doses of DDT. They also brought out more poisons from the same chemical family: aldrin, chlordane. The aim was nothing less than the control of nature, and pesticide manufacturers believed that control was within their grasp. The annual introduction of new pesticides rose from the very first product, DDT, in 1940, to great waves of chemical invention in the 1960s and 1970s. In those decades, dozens of new herbicides and insecticides were brought to market each year. *Heliothis* became one of the most heavily sprayed species in what amounted to a biological world war. Through it all, the moths clung to the cotton.

At the moment, most farmers in the Cotton Belt are spraying their fields with pyrethroids, which are sold under brand names as optimistic as ever: two of the more popular are Scout and Karate. When these pyrethroids were first introduced, they improved cotton yields by about a quarter, sometimes as much as a third. But in the year 1980, there were reports from California's Imperial Valley of moths with fifty-fold resistance to pyrethroids. This resistance movement spread like the others.

"Right now *Heliothis* is throwing the cotton industry in Louisiana in a panic," says Bruce Black, an entomologist at American Cyanamid, the pesticide giant, which is based in Princeton. "The moths have become almost absolutely resistant to all pesticides, from your cyclodienes to your organophosphates to your carbamates, and most of your pyrethroids. And these pyrethroids are literally the last bastion to prevent the cotton industry from collapsing. In Louisiana, in the fields, they have insects that are two-hundred-fold resistant to pyrethroids. Farmers are saying they can't grow a crop next year. The bugs will wipe them out."

Martin Taylor has a grant from American Cyanamid to watch *Heliothis* evolve.

WHEN HE LECTURES about *Heliothis*, Taylor usually begins with a transparency on which he has printed, in tall, spindly, Victorian characters:

THE VARIATION OF ANIMALS AND PLANTS UNDER DOMESTICATION

with special reference to resistance
of insects to pesticides

By Charles Darwin, MA, FRS, &c.

"It's an extraordinarily potent example of evolution going on under our eyes," Taylor says. "Visible evolution."

A pesticide applies selection pressure as surely as a drought or flood. The poison selects *against* traits that make a species vulnerable to it, because the individuals that are most vulnerable are the ones that die first. The poison selects *for* any trait that makes the species less vulnerable, because the least vulnerable are the ones that survive longest and leave the most offspring. In this way the invention of pesticides in the twentieth century has driven waves of evolution in insects all over the planet. *Heliothis* is only one case in hundreds. Flying scale insects have evolved strong resistance to buquinolate in only six generations. Nematode worms in the guts of sheep have evolved strong resistance to thiabendazole in three generations. Sheep ticks have evolved strong resistance to HCH–dieldrin in just two generations. In 1967, a distinguished entomologist announced in *Scientific American* the discovery of a "resistance-proof" family of insecticides. The poisons were variants of some of the insects' hormones. How could insects escape their own hormones? Yet within five years, flies had evolved one-hundred-fold resistance.

"This seemed to surprise people," says Taylor. "It would not have surprised an evolutionary biologist. But it surprised pesticide sprayers and the manufacturers of chemical compounds endlessly."

"If you look through the literature," says Linda Hall of Cornell University, who specializes in the study of pesticide resistance, "you'll find people saying, 'Resistance will not develop for pyrethroids.' That was incredibly naive. Almost anything you give an insect, almost any way you find to kill it, it will find a way not to be killed. That's the whole bit about evolution: no matter what you choose for your killing method, it will find a way not to be killed. Yet people from various companies were standing up at American Chemical Society meetings and saying, 'Insects should not develop resistance to pyrethroids.' I don't know," she says. "I don't understand it at all."

When evolutionists study these worldwide resistance movements, they see four classes of adaptations arising, because an insect under attack has four possible routes to survival.

First, it can simply dodge. Strains of malarial mosquitoes in Africa used to fly into a hut, sting someone, and then land on the hut wall to digest their meals. In the 1950s and 1960s health workers began spraying hut walls with DDT. Unfortunately in every village there were always a few mosquitoes that would fly in through the window, bite, and fly right back out. Millions of mosquitoes died, but these few survived and multiplied. Within a short time almost all of the mosquitoes in the villages were hit-and-run mosquitoes.

Second, if an insect cannot dodge, it can evolve a way to keep the poison from getting under its cuticle. Some diamondback moths, if they land on a leaf that is tainted with pyrethroids, will fly off and leave their poisoned legs behind, an adaptive trick known as "leg-drop."

Third, if the insect can't keep the poison out, it may evolve an antidote. A mosquito species called *Culex pipiens* can now survive massive doses of organophosphate insecticides. The mosquitoes actually *digest* the poison, using a suite of enzymes known as esterases. The genes that make these esterases are known as alleles B1 and B2. Many strains of *Culex pipiens* now carry as many as 250 copies of the B1 allele and 60 copies of B2.

Because these genes are virtually identical, letter by letter, from continent to continent, it seems likely that they come from a single lucky mosquito. The mutant, the founder of this particular resistance movement, is thought to have lived in the 1960s, somewhere in Africa or Asia. Its descendants apparently hitched rides around the world in airplanes. The genes first appeared in Californian mosquitoes in 1984, in Italian mosquitoes in 1985, and in French mosquitoes in 1986.

Finally, if the insect can't evolve an antidote, it can sometimes find an internal dodge. The poison has a target somewhere inside the insect's body. The insect can shrink this target, or move it, or lose it. Of the four types of adaptations, the four survival strategies, this is the hardest for evolution to bring off—but Taylor thinks this is how *Heliothis* is evolving now.

"It always seems amazing to me that evolutionists pay so little attention to this kind of thing," says Taylor. "And that cotton growers are having to deal with these pests in the very states whose legislatures are so hostile to the theory of evolution. Because it is evolution itself they are struggling against in their fields each season. These people are trying to ban the teaching of evolution while their own cotton crops are failing because of evolution. How can you be a Creationist farmer any more?"

Heliothis can evolve resistance to pyrethroids in the course of a single growing season. In Arkansas in May 1987 only about 6 percent of the moths survived a certain fixed dose of the poison. But by that September, several moth generations later, 61 percent survived that same dose. The same rapid evolution has been observed in cotton fields in Louisiana, Oklahoma, Texas, and Mississippi.

Years ago, when DDT resistance first appeared, geneticists studied the problem in the laboratory with house flies. Flies that survived doses of DDT often carried a certain mutant gene on the third chromosome, a mutant that came to be called *kdr*, for knockdown resistance. This single gene conferred resistance to DDT and all its variants.

Today, every postmodern, well-equipped house fly carries not only *kdr* but also a mutant gene called *pen*, which reduces its uptake of insecticides. On the fly's fourth chromosome it carries a mutant gene called *dld-r*, which gives it resistance to dieldrin and dieldrin's family of poisons. On its second chromosome it carries a mutant gene known as *AChE-R*, which protects it from organophosphates and carbamates.

Flies were astonishingly quick to evolve resistance to pyrethroids, and many investigators think they coped so well because they had already evolved in the right direction with DDT. The same *kdr* that fought DDT seems to have fought pyrethroids too. If a lucky fly's parents both carry *kdr*, so that it inherits one copy from the mother and one copy from the father, it will often display one-thousand-fold resistance to pyrethroids.

According to current thinking, DDT and pyrethroids both attack the same target in the fly: microscopic doorways in the membranes of the fly's nerve cells. These doorways, sodium channels, open and close to allow nerve signals to pass through the cell. Both DDT and pyrethroids are thought to jam the channels open, triggering repetitive, uncontrolled discharges in the nerve cells. If enough of the channels get locked open, a fly goes into convulsions, then paralysis, and dies.

The structure of the sodium channel is almost identical in animals as far apart on the evolutionary tree as flies, eels, rats, and human beings. This suggests that the structure evolved far back in evolutionary time, before the split in the tree between invertebrates and vertebrates. In the sodium channel, then, the poison is attacking a structure that is old, vital, and universally fixed in its design throughout the animal kingdom. One would think it would be extremely hard for a fly to change such a venerable design. But flies have done just that. They have modified the genes that make the channels in a way that somehow saves the flies from the jamming action of the poison.

The sodium channel gene has been completely sequenced in *Drosophila melanogaster*. That is, all of the invisible characters that make up the gene have been deciphered and published. So Taylor and his advisor, Marty Kreitman, decided, on a hunch, to go into the DNA of the moth and look for the same gene. Although the *Heliothis* genome is unusually big for an insect—it is bigger than a chicken's, approaching the size of a human being's, a billion letters long—Taylor knew exactly how to go about finding the gene. With the new molecular techniques the procedure is reasonably easy to do.

The *Drosophila* gene goes ATCGAGAAGTACTTCGTGT . . . and so on. Taylor went to a small machine along the wall of the laboratory, a DNA synthesizer, which is standard equipment for a molecular evolutionist. Standing at this synthesizer, as nonchalantly as if he were typing on a keyboard, he instructed the machine to assemble the matching sequence of letters: a fragment of artificial DNA. Then he ground up bits and pieces of *Heliothis* moths in a mortar and pestle, extracted the moths' DNA, and mixed in the artificial DNA.

If the moth did carry the same gene as the fly, Taylor's DNA fragment would find it, and bind to it. The small strand of artificial DNA would stick to the long strand of moth DNA. To see if they did stick together, he bathed all of the DNA in a special enzyme solution, which chopped the strands into pieces, and he sifted through the pieces.

("It's very robust stuff, DNA," Taylor says. "You can do all this grinding and it won't break. The only risks to it are from enzymes.")

After months of work, Taylor found a fragment in the genome of the moth that matched part of the gene in the fly. The moth gene went ATCGAGAAGTACTTCGTGT ... and so on for 184 of Darwin's invisible characters. Almost two hundred letters, and they differed by only a single letter between *Heliothis* and *Drosophila*. The moth carries almost exactly the same gene as the fly.

Now Taylor is searching through the moth's sodium channel gene for changes that might protect the moth from pyrethroids. Which letters have flipped along the length of the DNA, and saved the day for *Heliothis*? This part of the search is much more laborious. It may take as little as a single letter-flip, a single point substitution, as it is called, to make the moth resist the poison. And the whole gene consists of many thousands of letters.

Yet the pesticide manufacturers once thought they would wipe out pests once and for all with pyrethroids. "People don't like to have their categories threatened," says Taylor. "They like to think, That's a moth, and That's a fly. Fixed categories. They don't like to think of lots of hybridizing and change all the time in every line of moths and flies, lots of evolution in action all the time around them.

"Even people who think they understand Darwin tend not to think about this, because they are schooled in Darwin's gradualism.

"It is hard enough to control pests even if we know what is going on. But you can't control them if you don't realize the target you are aiming at can *move*."

BECAUSE WE DO not think this way, we make the same tactical errors again and again. Resistance movements are not only out in the cotton fields. They are also closer to home. During the past fifty years while we have been throwing poisons at the pests in our fields, we have also been throwing larger and larger quantities of chemicals at the pests within our own bodies. Here too scientists are now watching evolution in action.

Western hospitals started to use antibiotics regularly in the 1950s, and resistance appeared within a year or two. One out of three patients in every Western hospital is now on antibiotics, and antibiotic resistance is increasing so rapidly that many physicians are calling it a global epidemic.

This kind of resistance tends to follow the same cycle as pesticide resistance: big companies, big medicine, blanket treatment, followed by almost immediate disappointment.

"It's generally the case that these chemical companies are not familiar with evolution," says Martin Taylor.

In the U.S. the Centers for Disease Control and Prevention (C.D.C.P.) in Atlanta has instituted a national surveillance for drug resistance. "When these new drug-resistance strains become endemic in hospitals," one doctor told the editor of *Science* recently, "you will be safer staying home than going to a hospital unless you have a truly dread disease."

It is easy to start a resistance movement in the most common bacteria in the human gut, *E. coli*. First one establishes a colony of *E. coli* in a Petri dish. The bacteria multiply so rapidly that a single microscopic cell can grow into a visible pile of ten million *E. coli* between morning and mid-afternoon. To the eye, ten million *E. coli* look like a tiny heap of salt.

Next, one doses the colony with an antibiotic. As fast as it grew, the colony disappears. Only a few cells in the colony have survived—the two or three cells that carry a rare resistance gene for that antibiotic. These several survivors multiply and pass their successful gene to their descendants. Soon there is a new colony in the dish, a colony in which virtually every member cell is resistant to the antibiotic.

"Darwin would have loved to watch that simple experiment," says one molecular evolutionist. "That's exactly what he said natural selection is. And it can happen in a day or two."

Bruce Levin, a microbiologist who is now at Emory University, in Atlanta, once teamed up with several colleagues to watch the evolution of *E. coli* in his own gut. Every few days when he went to the john he would take a sample. (A single wipe of toilet paper comes away with as many as two trillion individuals of the *Bacteroides* species, twenty billion individual enterobacteria, and dozens of other species that have never been named by science.)

The investigators followed Levin's *E. coli* for almost a year. They found that the ecology of his gut was hectic, eclectic, and tumultuous. Strains of *E. coli* kept appearing and disappearing. In the course of the experiment, the microbiologists identified a total of fifty-three different strains, all but two of which went rapidly extinct. These strains of bacteria were apparently colonizing Levin like birds in the Galápagos. The strains were flying in on the food he ate and on every touch from his wife, their two children, their dog and cat, a group that a commentator in *Nature* dubbed "the Levin Archipelago."

In these bacterial cells the DNA is not locked away in a walled nucleus. Instead it floats free in a long necklace of DNA, a circle of about ten thousand genes. Other smaller necklaces float about within the cell too, rings of DNA called plasmids. A typical *E. coli* individual holds two or three plasmids. Dozens of these plasmids can move among different species of bacteria like secret battle codes, written in Darwin's invisible ink. Some of their genes can actually jump from the plasmid and insert themselves into the main necklace, or eject themselves completely from the cell and enter another bacterial cell, like letters without envelopes. When bacteria are under stress—for instance, when their human host takes an antibiotic—the cells use their jumping genes to pass resistance genes back and forth at a great rate.

"We did a study," Levin says. "How fast can evolution proceed in a human? My wife took ampicillin. I took erythromycin. Within a few days, we were both dominated by resistant bacteria. Not only was tetracycline resistance coming up, but also streptomycin, kanamycin, carbenicillin—our bacteria were going from almost nothing to multiple resistance in an amazingly short amount of time.

"If you did it in a test tube you wouldn't be surprised. But to take a couple of pills and see it happen inside you is really kind of awesome. It gives you an eery feeling: when we talk about natural selection, we're not talking about eons here. It's not just dead dinosaurs."

Resistance is now rising in gonorrhea, streptococcus, tuberculosis, salmonella. In the United States, the incidence of the bacterium *Neisseria gonorrhoeae* with resistance to penicillin more than tripled between the years 1988 and 1990. In 1990 an outbreak of fatal dysentery struck Burundi, and the microbes were resistant to every single oral antibiotic in the country.

Local resistance movements like these can get carried around the world and plunge millions of people into epidemics. In principle this is nothing new. Measles traveled the caravan routes of the Roman Empire from A.D. 165 to A.D. 180. Smallpox followed the same routes in A.D. 251 to 266, and one in three people along the caravan routes died. But today the caravan routes are mostly airborne, and Concordes are faster than camels. A new virus or bacterium can circle the earth in a matter of days.

"In 1941, 10,000 units of penicillin administered four times a day for four days cured patients of pneumococcal pneumonia," writes the physician Harold Neu of Columbia University. Today a patient could receive twenty-four million units of penicillin a day and still die of the disease. "Bacteria are cleverer than men."

Not long ago, reading the headlines in a front-page newspaper story about drug resistance, a molecular evolutionist stopped at the word *inordinately*. The headline said that bacteria in some hospitals are "inordinately" resistant to penicillin. "Well," he said, "they're not inordinately resistant to it. They're completely resistant to it. There might as well not be penicillin there."

"Considering the shortness of the time span, the number of mechanistically different countermoves that bacteria have invented against antibacterial agents is amazing," writes Alexander Tomasz of Rockefeller University. These cells have evolved an appalling arsenal of weapons against penicillin and its large family of antibiotics. They have evolved anti-antibiotic enzymes to match every antibiotic that is being thrown at them. All these chemical weapons and counterweapons, says Tomasz, "match one another as defensive and offensive weapons match in classical warfare: shield against the arrow, bazooka against the tank." There are now drugs designed to attack the bacteria's resistance to antibiotics: anti-anti-antibiotics. And much of this high-technology war has evolved in the lifetime of doctors now in practice. "The amazing degree of variation" in some kinds of resistance, writes Tomasz, suggests that these resistance traits are "continuing to evolve under our very eyes."

As with pesticides, investigators are now tracking the evolution of antibiotic resistance at the level of DNA. The core of anti-tuberculosis regimens is a drug known as isoniazid. Recently investigators looked at strains of *Mycobacterium tuberculosis* isolated from two patients. They found that in each strain, the bacteria had dropped from its chromosome a gene called *kat*G, which codes for the production of two enzymes, catalase and peroxidase. In the laboratory the investigators isolated a strain of bacteria that lacked this gene. The strain produced very little of these two enzymes, and was resistant to isoniazid. Into this strain, the investigators inserted the missing gene *kat*G. Instantly the strain started manufacturing the two enzymes, but it was now killed by isoniazid.

Apparently the cells had paid a price to defend themselves from the drug. They had made an evolutionary trade-off, giving up part of their own adaptive equipment for the sake of survival. The bacillus got rid of an Achilles' heel by evolving a heel-less foot.

Each year about eight million people are colonized by this adaptable bacillus, *Mycobacterium tuberculosis*. About one in three people on the planet already carry it, and each of them has about one chance in ten of developing the symptoms. In the developing world the disease accounts for almost 7 percent of all deaths. In the United States in the 1980s tuberculosis had been in retreat for a century. Doctors and the directors of public health programs thought of it as a defeated disease. They looked away. Between 1985 and 1992 tuberculosis increased in incidence by nearly 20 percent. The number of cases among children born in the United States under the age of five rose 30 percent just between the years 1987 and 1990. As two physicians observe in a review of the disease and its resurgence, "The principal risk behavior for acquiring TB infection is breathing."

PHYSICIANS CAN WATCH the human immune system attack invaders, and they can watch the bacteria and viruses dodge and twist. F. MacFarlane Burnet, who won the Nobel Prize for his work in immunology, called this "evolution made visible." Teams of investigators are now watching the evolution of AIDS in the living bodies of individual patients. Intensive studies of this kind have been conducted in England, the United States, and Africa. Viruses are the first organisms on the planet whose genetic sequences have been read and published from beginning to end. The complete nucleotide sequence of one strain of the AIDS virus HIV-I is 9,749 base pairs long. But this sequence does not stand still, because the virus has no proofreaders. When investigators take a series of samples from an individual patient they see rapid evolution. Individual letters in the sequence change, clusters of letters change, whole chunks of DNA disappear while other chunks insert themselves in new places along the strand. A human body with AIDS is like an entire Galápagos archipelago: it harbors an increasingly diverse group of viruses after the first one has invaded it. The first virus particle to invade evolves into a swarm of variant strains.

In the AIDS virus the gene known as *env* is the one that evolves fastest. *Env* codes for the envelope of the virus, which is what the human immune system tries to grasp and destroy. The *env* gene changes about a million times faster than the normal mutation rate of its host, the human body. In this way, according to present thinking, it keeps eluding the grasp of the immune system. In a sense, the weapon of the virus is variation itself.

The influenza virus also evolves rapidly. The sequence of the human flu virus changes at a rate of more than two letters a year. It can evolve even faster if it happens to meet up with a virus that infects horses, pigs, or seagulls. Very often the strains meet when they both happen to infect the body of the same pig. There they not only give the pig the flu; they also swap genes. They hybridize, and the new virus is sometimes changed enough to evade the attacks of the human immune system and sweep the world.

So far, despite all our plagues, we have been lucky. In principle, a random mutation or a hybridization event could someday create a virus that combines the airborne, infectious qualities of flu and the deadly, long-latency, slow-killing qualities of AIDS. It hasn't happened yet, but there is nothing in Darwin's process to prevent it, and the larger the pool of human beings on the planet, the more viruses are jumping in. "Our only real competition for domination of the planet remains the viruses," the microbiologist Joshua Lederberg once said. "The survival of humanity," he added, "is not preordained."

Resistance movements can spring up and overwhelm us even within the lines of our own cells. A cell that turns cancerous is a cell that has escaped the molecular restraints that keep most of our cells from multiplying out of control. When doctors hit a colony of these cells with drugs, radiation, or heat, a few of the cells may resist the attack.

"With cancer, when you are in chemotherapy, you get resistance," says Bruce Levin. "You are seeing the very same problem in yourself." Most of the cells of a tumor are typically killed by the first doses of chemicals, but those that survive can proliferate. "The rapidly growing cells—it behooves them to avoid being killed," says Levin. "It's no different from putting streptomycin into a bacterial culture. Talk about evolution in action! You're seeing evolution at home. Of course, that's the last thing on your mind."

ALL THIS IS a simple corollary to Darwinian law. Wherever we aim at a species point-blank, for whatever reason, we drive its evolution, often in the opposite direction from what we ourselves desire. The law holds, whatever our reason for shooting at a species, and whether the species is submicroscopic or gigantic.

In the late 1970s and early 1980s, between 10 percent and 20 percent of all the elephants in the wild were being killed each year. At that rate wild elephants would have gone extinct by the end of the century. This was an intense selection event.

For poachers, elephants with big tusks were prime targets. Elephants with small tusks were more likely to be passed over, and those with no tusks were not shot. In effect, though no one realized it at the time, African elephants in places where poaching was rife were under enormous selection pressure for tusklessness. And in fact, elephant watchers in the most heavily poached areas began noticing more and more tuskless elephants in the wild. Andrew Dobson, an ecologist at Princeton, has compiled graphs of this trend, tracing the evolution of tusklessness in five African wildlife preserves, Amboseli, Mikumi, Tsavo East, Tsavo West, and Queen Elizabeth. In Amboseli, where the elephants are relatively safe, the proportion of tuskless female elephants is small, just a few percent. But in Mikumi, a park where the elephants are heavily poached, tusklessness is rising. The longer each generation lives the fewer tusks the elephants carry. Among females aged five through ten, about 10 percent are tuskless; among females aged thirty to thirty-five, about 50 percent are tuskless.

Male elephants use their tusks to fight one another over females. When most males are tusked, a male without them is like a knight without a lance. But where fewer and fewer males sport tusks, the untusked male has a fighting chance of getting a harem and passing on his tuskless genes. He is more and more fit. These are evolutionary changes, and no matter what happens in the future, the balance of genes will work itself out for many generations and many centuries—if elephants live that long.

This same sort of evolutionary change is also taking place in the world's seas. Most sport fishermen and commercial fisheries follow a basic rule: keep the big ones and throw back the little ones. That too is an evolutionary pressure. A net is a powerful agent of Darwinian selection. In one recent laboratory demonstration, investigators raised water fleas in aquarium tanks. Every four days they sieved the tanks with fine-meshed nets. In one set of tanks they threw back the little fleas and killed the big ones. In another set of tanks, they threw back the big fleas and killed the little ones. They kept this up for generations of water fleas, and they saw a dramatic evolutionary response. In the tanks where they culled the small fleas, the fleas began growing faster and delaying the age of first reproduction. Fleas that put all their energy and resources into growing fast had the best chance to escape the net. They saved the act of reproduction (which is expensive in terms of time and resources, even for a flea) until they were older, bigger, and safer.

But in the tanks where the big fleas were killed, evolution ran the opposite course. There the water fleas grew slowly and began reproducing when they were still small. For those fleas it was the one that stayed small longest that lived longest and passed on the most genes.

John Endler has seen the same kind of evolutionary responses among his guppies, both in the wild and in the laboratory. Some guppy eaters like their guppies big, and some like them small, and they too drive the evolution of their prey. The changes are predictable and rapid; they take about fifty guppy generations.

In the world's oceans, Norwegian cod, chinook salmon, Atlantic salmon, red snapper, and red porgy are getting smaller, very likely through the selection pressures of the net. Fishermen are not happy with the trend toward small fish, any more than elephant poachers are pleased with the trend toward tusklessness. But both resistance movements are direct results of Darwinian law.

"WORKING LATE?" asks one of the lab assistants.

"As usual," Taylor says. He is wearing two days' worth of black stubble, a scruffy black turtleneck, and a slightly disreputable shade of gray in his face.

"I don't know why you bother to keep an apartment," says the assistant. "You're always here. You might as well just string up a hammock."

"Might as well," he says.

It is past midnight, and Taylor is grinding away, but the moths are evolving much faster than he can keep up with them. They are growing more and more resistant in Australia, and also in the United States, mostly west of Alabama. Apparently there is more than one gene involved in their resistance movement. The sodium channel may be only the beginning of the story.

Meanwhile most cotton farmers are still spraying pesticides routinely, often before any pests show up, a practice known as "insurance spraying" or "spray and pray." So the selection pressures on *Heliothis* and the other pests in the cotton fields, including the notorious boll weevil, continue to be intense, and the pests keep evolving more and more resistance. As more and more developing countries turn from DDT to pyrethroids, they are likely to provoke precisely the same evolutionary events, the same evolution of *kdr* and the same control failure.

The chances of success of each new pesticide under development keep dropping, and the costs of development keep rising. One expert wonders whether we have already selected in pests the genes they need to evolve resistance "to practically any toxicant that may be used against them. The answer," he adds, fatalistically, "will be provided in time by the pests themselves."

Before human beings had heaped up a mountain of pesticides in the 1940s, when we were still in the foothills of this evolutionary adventure, farmers in the United States were losing about 7 percent of their crops to insects. During the blitz of the 1970s and 1980s the insects did not lose any ground. Instead they nearly doubled their share, to 13 percent. "Indeed," note the ecologists Robert May and Andrew Dobson, "the fraction of all crops lost to pests in the United States today has changed little from that in medieval Europe, where it was said that of every three grains grown, one was lost to pests . . . leaving one for next year's seed and one to eat."

May and Dobson believe this global evolutionary disaster "may help to show that evolution is not some scholarly abstraction, but rather is a reality that has undermined, and will continue to undermine, any control program that fails to take account of evolutionary processes."

Meanwhile, the ten least-wanted microbes in the world today, from enterobacteria to streptococcus, are now resistant to almost everything we can throw at them. Drug companies are bringing out only a few new drugs against them each year, a process that costs about $200 million per drug, and each new drug often takes as much as seven years to bring to market. Harold Neu of Columbia says it is now critical for doctors, patients, and drug companies to avoid all unnecessary use of antibiotics, for humans or animals, "because this selective pressure has been what has brought us to this crisis."

What we don't understand on either front is that the more pressure we put on our pests, the more we cause them to evolve around the pressure. The pressure is evolutionary pressure; what we fail to understand is evolution itself. Evolution is not just the Galápagos and not just out there beyond the windowpane buffeting the robin and the oak. Evolution is a very near thing. To us it is a terrible irony. Precisely where we wish to control the environment most tightly and possess it most completely we are powerless to do so, besieged and beleaguered by resistance movements that seem to spring up faster the more we lop them off, like the heads of Hydra. The harder we fight these resistance movements, the harder and faster they evolve before our eyes—precisely because it is our effort at control that is driving their evolution. What we call control is to them merely a change in the environment, just another change in an endless series of changes— and they are superbly placed and designed to keep up with such changes. As long as we keep up the pressure indiscriminately they will continue to rise in plagues against us, like the frogs that came up on the land of Egypt, or the dust of the earth that became lice throughout all the land of Egypt.

"RIGHT," TAYLOR SAYS, in the voice of midnight, as he lifts another moth from its crater of ice and holds it up with his tweezers. "Hey, what's your trick?"

He drops it in the bottom of a clear plastic vial. It looks out of place lying there, surrounded by the whirling centrifuges and chirping Geiger counters of the laboratory, like an African mask on the wall of a sleek modern art gallery. Taylor fusses with his preparations. "It's very—hmmmm—tedious," he mutters. He picks up a glass rod, to grind a bit of the moth in the bottom of the vial.

"Not yet resistant to the pestle," he says.

Chapter 19

A Partner in the Process

Creation is not an act but a process; it did not happen five or six thousand years ago but is going on before our eyes. Man is not compelled to be a mere spectator; he may become an assistant, a collaborator, a partner in the process of creation.

— THEODOSIUS DOBZHANSKY,
Changing Man

In the last decade, Rosemary and Peter have witnessed two of the most extreme years of the century in these islands: the wettest and the driest. In the wettest, more rain fell in a single day than falls in a normal wet year. In the driest, not a single drop of rain fell. As Peter says, you can't get a drier year than that.

At any other time, the flood and the drought would have been seen as acts of God or freaks of nature. But today, when the Grants think about those Malthusian years, they wonder.

"The idea that organisms evolve was transformed during the last century from conjecture to fact," they write in *Noticias de Galápagos*, the journal of the Charles Darwin Research Station. "As the present century draws to a close, we are experiencing another transformation. The conjecture that the world's temperature is gradually rising has become widely accepted as a demonstrated fact."

Most earth scientists now agree that during the last one hundred years, the planet's surface temperature has risen, with hesitations and reversals, by about half a degree centigrade. This now-notorious global warming began around the time of Darwin's death, in the 1880s, and it became strongly marked by the close of the 1980s, which was, as the Grants write, "undoubtedly the warmest decade of the century."

Meanwhile the amount of carbon dioxide in the atmosphere has also risen, with fewer hesitations and no reversals. Other gases have accumulated too. All of them are by-products of human industry and agriculture: carbon dioxide, carbon monoxide, and nitric oxide are billowing invisibly from our fires, smokestacks, and exhaust pipes; methane from our vast fields of cattle, sheep, and rice. The gases trap heat, which is why they are called greenhouse gases, and most earth scientists expect them to trap more and more heat in the next century.

The forecasts are highly uncertain; the crystal ball is cloudy. "Nevertheless," the Grants write, "we should be thinking about the implications of global warming for Galápagos."

Global warming is of special interest here because in these islands the round of the seasons is driven by ocean currents. Half the year the archipelago is bathed in cool waters, the other half in warm. The cool waters come from the South Equatorial current, and the warm waters from the North Equatorial current. The difference between these currents is often 10° C, sometimes as much as 20°, which is more than enough to bring distinct seasons to the islands.

If it were not for these alternating currents the islands would have no seasons at all, since they lie exactly on the equator. Nor would they carry their bizarre flora and fauna. It is because they are a meeting place of waters from north and south that the passenger list of the islands is so diverse, including not only tropical lizards but also fur seals, not only tropical flamingos but also penguins—the only penguins on the equator.

A warming of even half a degree centigrade can cause a change in global circulation patterns, and make trade winds and ocean currents veer away on new tracks. Because they depend on winds and currents for their very seasons, the Galápagos Islands are particularly vulnerable to changes like these. That is why the finch watchers on Daphne Major always put quotes around the word "*typical*" when they describe a typical wet season or dry season. The currents are so variable that no two years are alike.

What is more, the Galápagos lie near one of the key pressure points in the global circulation system: the birthplace of El Niño. The arrival of El Niño every few years has the effect of heightening and lengthening the contrast between the warm bath and the cool bath.

Though no two years are alike, the Niño years are more different than the others. Each Niño turns life upside down coming and going.

If the winds and currents and hence the seasons were not so variable, the Galápagos finches would not need such variable beaks. Indeed it must have been a freak of these variable currents that first swept the finches to the islands. These winds and currents helped make Darwin's finches what they are and they are still shaping the finches today.

It is no exaggeration to say that a lasting change in the ocean currents—especially a change in the intensity or the frequency of El Niño—would change the course of evolution in Darwin's islands. And the islands sit in a spot where even a slight global warming could make an enormous and early difference.

A few years ago, when they got back from Daphne and began their annual catch-up with the world's news, Rosemary and Peter were interested to read a short article that had appeared in the journal *Science*. The article was written by a climatologist at the U.S. National Oceanic and Atmospheric Administration, Andrew Bakun, who described the implications of his study as "uncertain but potentially dramatic."

If present thinking about global warming is correct, Bakun argued, and Earth's atmosphere is growing warmer, then the surface of the planet should be responding unequally. The land areas should be warming faster than the sea. (In the seas, cold water is always being churned up from below.)

If the continents are warming faster than the seas around them, Bakun reasoned, then the contrast in temperature between coastal lands and coastal waters should be increasing. One consequence might be an acceleration of offshore winds, because such winds are driven precisely by this difference in temperature between the coastland and the sea.

Bakun collected wind-stress records for the northern and southern coasts of Peru, and also for California, the Iberian Peninsula, and Morocco. He found that along every one of these coastlines, wind-stress has risen significantly since mid-century. The waters off Peru, the birthplace of El Niño, present an extreme case, according to Bakun; in the middle of this extreme case, of course, sit the Galápagos Islands.

So the freak weather the Grants have seen in the past ten years may have been something more than chance. It may, just possibly, have

been influenced by global warming, and if so it may be only the beginning.

IN THE *Origin* Darwin asks readers to imagine, hypothetically, "the
case of a country undergoing some slight physical change, for instance,
of climate." If the country were part of a continent, its borders might
be flooded with immigrants, Darwin writes, with endless trains of evolutionary consequences. Even on a lonely island the inhabitants
would adapt to the altered conditions, "and natural selection would
have free scope for the work of improvement."

No islands are too remote for these experiments, not even the
Galápagos. In fact, with Darwin's islands once again nature seems to
have arranged a particularly dramatic case, a demonstration of the
power of slight physical changes to push Darwin's process in surprising
directions. A rise in global temperature of half a degree centigrade is
probably less than what Darwin has in mind when he speaks in this
passage of "a slight physical change." A rise in carbon-dioxide concen-
tration of one part per million per year might have seemed to Darwin
almost too slight to consider, too small to propel his process. But
changes even as small as these can propagate through Earth's climate
system and through Darwin's "web of complex relations" until they
literally change the face of the world.

The Grants' argument about the fission or fusion of Darwin's
finches, for instance, depends on the islands' climate staying more or
less as they have seen it in the last twenty years. That is, when they
project the fate of Darwin's finches, Rosemary and Peter have always
assumed there will be no net change in the seasonal cycle of the archipelago. They have assumed that the pendulum of ocean currents will
go on swinging annually but erratically between wet seasons and dry
seasons. But this is no longer a safe assumption, the Grants write:
"Global warming alters the argument."

If global warming did give birth to the Wild Child of 1982, then
more warming, if it comes, may well bring more Niños like that one.
The Niño of 1982 has often been described as the strongest of this
century. Some climatologists now believe it was the strongest of many
centuries, perhaps the strongest in the second half of this millennium.

Before that flood year, on Daphne Major, Darwinian selection was
keeping Darwin's finches apart and distinct. After the flood, selection
pressures on Daphne began forcing the birds together. If global warm-

ing does bring more extreme Niños, then the selection pressures on the islands may take a very long time to return to what they were before the flood. The conditions on the islands may not return to what they were for many decades, the Grants speculate, "perhaps a century, in which the prospect of three species fusing into a single population becomes more likely."

The small, medium, and large ground finches, those bicycle-pump productions, are extraordinarily responsive to changes in the weather of the islands. If Niños were to begin to come harder and faster in the next century, then it would take only about two hundred years for the finches to fuse together and undo all the evolutionary work that has carved them apart.

On the other hand, it might not take Darwin's process very long to separate the birds again: to turn a group of *fuliginosa* into a *fortis*, or a *fortis* into a *magnirostris*. Trevor Price has calculated that it would take about twenty selection events as intense as the drought of 1977 to turn a *fortis* into a *magnirostris*. And if the starting point were not Daphne but one of the islands where *fortis* is larger, then the change would take only a dozen droughts. "Trevor worked out that in a relatively small time you could get from A to B," says Peter Grant. "I hadn't, when we started this work, even thought that would be possible."

In other words, in the present climate of the Galápagos, it would take only a thousand years of not unlikely weather to create a new species of Darwin's finches on the islands. And if the climate were to change and inflict a series of grim droughts or floods at just the right intervals, without missing a beat, it could create a new species in a single century.

FOR THE MOMENT the link between Galápagos weather and global warming remains speculative. But the case suggests the kinds of Rube Goldberg–like and ultimately unpredictable local events that global warming may inflict in our lifetimes, even on some of the most isolated islands in the world.

And our power to drive Darwin's process, like the power of the process itself, is not hypothetical. The industrial revolution was changing environments and, with them, the course of evolution even before Darwin published the *Origin*. That is the message of the single best-known evolution watch in history.

In 1848, a lepidopterist in Manchester by the name of R. S. Edleston put a pin through a moth, a rare form of the species *Biston betularia*. The normal form of the moth was whitish in color, with a peppering of fine black lines and spots. But Edleston's specimen was almost as black as coal, from which it got its scientific name, *carbonaria*.

Collections feed on rarities. In the second half of Darwin's century the black moth was prized by every butterfly and moth hunter in the British Isles. It was so ardently sought after that evolutionists in our century have been able to use Victorian lepidopterists' records and collections to trace the spread of the black mutant across England. In 1860 a specimen of *carbonaria* was netted in Cheshire, in 1861 in Yorkshire, in 1870 in Westmorland, in 1878 in Staffordshire, in 1897 in London. Soon after the first sighting the black mutant became more common in each of these places, until the white form at last grew rarer than the black.

Black mutants conquered the Continent too. In 1867 a pair of them were caught copulating on an elm tree in the Netherlands, in the province of North Brabant. In 1884 the black mutants were reported in Hanover, and in 1888 in Thuringia. From there they seem to have made their way up the Rhine Valley.

The black mutants swept up through the moth populations wherever the air was black with the soot of the industrial revolution. Their numbers did not rise in rural parts of Cornwall, Scotland, and Wales. In rural Kent, Darwin's adopted county, the black form of the moth was not recorded during his lifetime; but by the middle of this century, nine out of ten *Biston betularia* were black in Bromley, and seven out of ten in Maidstone.

Manchester, of course, was one of the grimy hubs of the industrial revolution. At about the time that Edleston pinned his *carbonaria* there, the novelist Elizabeth Gaskell was describing a family's first sight of the city as they rode in on a train: "Quickly they were whirled over long, straight, hopeless streets of regularly built houses, all small and of brick. Here and there a great oblong, many-windowed factory stood up, like a hen among her chickens, puffing out black 'unparliamentary' smoke, and sufficiently accounting for the cloud which Margaret had taken to foretell rain."

That black cloud was "unparliamentary" because air-pollution laws had been passed even then. But the laws had no bite. Soot was blackening the trees around Manchester, and around every other dark mill town in England. In the twentieth century, experiments by

H. B. D. Kettlewell of Oxford showed that this soot made a mortal difference to moths that landed on the trees. Kettlewell filmed hedge sparrows, spotted flycatchers, yellowhammers, robins, thrushes, and nuthatches eating moths. On the pale bark of rural birches and beeches the birds were quicker to spot the black moths. But on blackened bark around cities, the birds were quicker to pick off the white moths.

The difference between the black form and the white is a single gene. Before the industrial revolution the black form was under strong negative selection pressure and the mutation stayed rare, except in forests with mostly black-barked trees. Factories reversed the selection pressure because the rare moths looked like soot themselves. The case of the peppered moth gave evolutionists one of their first inklings of the speed of Darwin's process. Suppose in the year they were first reported, 1848, the incidence of black mutants around Manchester was one in a hundred (almost certainly a generous estimate). Fifty years later, in 1898, ninety-nine out of a hundred were black. From these figures, the British evolutionist Haldane calculated that the advantage of each black moth over each white moth must have been running as high as 50 percent throughout the second half of Darwin's century. That is, generation after generation, a black moth was 50 percent more likely than a white one to pass on its genes.

In the middle of the twentieth century, however, Britain enacted strong clean-air legislation. The city air began to clear, and so did the bark of the trees outside the cities. By 1966, Manchester, one British *Biston* expert has written, "was still distinctly satanic in appearance but was being cleaned and rebuilt." As environmental legislation was passed in the rest of Western Europe, more and more white moths began cropping up again. In West Kirby, in northwest England, where nine out of ten *Biston* were black in 1959, their incidence was down to five out of ten in 1985, and fewer than three out of ten in 1989. In the Netherlands, in the province of North Brabant, where the first two black mutants on the Continent were seen fomenting revolution in 1867, the frequency of the black mutant dropped from seven in ten to fewer than one in ten.

Today *carbonaria* is declining rapidly virtually everywhere in Britain. So are dark forms of ladybirds and dozens of other British insects whose bodies had blackened with the Industrial Revolution. Evolution has reversed itself. Finch watchers have seen it reverse from drought to flood and flood to drought, and moth watchers have now seen it reverse from the industrial to the postindustrial age. At present rates, *carbonaria* will be as rare as it was before the Industrial Revolution by about the year 2010.

CARBON DIOXIDE IS really no more esoteric than soot. It is a product of the same process of combustion, and it comes out of the same smokestacks. But the gas is invisible, and its influence is global rather than local. If it does have the impact on our next century that earth scientists are now warning us to expect, then this global effect makes it far more important than soot: the rise of this gas will prove to be the single most important physical change to take place on our planet in a long time. If present thinking is correct, temperatures in the next hundred years may be higher than they have been in the last several million years, and the change may come as much as ten times as fast as it did in those millions of years, a shocking evolutionary experiment.

Meanwhile, genetic engineers are quite consciously manipulating and accelerating evolution. Some genetic engineers go so far as to call their work the Generation of Diversity: G.O.D. And this is precisely what they are doing: generating diversity. Their laboratories of evolution are only a few years old, but they hope to grow more prolific than the Galápagos. They are creating new corn and rice, new bacteria, new guinea pigs, a patented Harvard Mouse. With the same tools and techniques they could start re-engineering human beings, if we let them.

What the genetic engineers and their G.O.D. are doing now can make your hair stand on end. Not long ago, two engineers reported in *Nature* that they had built two new molecules that fit in the spiral of DNA: new rungs in the twisted ladder, new steps in the spiral stair. That is, beside Darwin's invisible characters, T, A, C, and G, they had added two new letters to the alphabet of life, which they called K and X. Writing in the same issue of *Nature*, a colleague of theirs wondered exuberantly how many more letters we can discover (twelve seems possible) and what new messages and new creatures we can compose with this new alphabet.

Our wild acceleration of life's central process makes the study of evolution excruciatingly timely. We are changing the environments of life faster and faster, and we are changing their genetic machinery faster and faster. All of this is G.O.D., the Generation of Diversity—as well as the Generation of Destruction. Greenhouse and field studies are being conducted of the forced evolution of oak seedlings, wheat, corn, butterflies, moths, and aphids due to the extra carbon dioxide our species is adding to the air. Microcosm and open-sea studies are being conducted of the accelerated evolution of Antarctic plankton under the ozone hole. A study of plankton at Anvers Island in the Antarctic in 1987 and 1988 found that UV-B radiation is penetrating as deep as 20 meters down into the sea beneath the ozone hole, so there is potential for a great deal of damage, and also of evolutionary response. A recent study found that as the ozone hole sweeps by overhead the higher levels of UV-B cut the numbers of plankton in the water by 6 to 12 percent, for as long as they were under the hole. Fortunately the rays did not hit all species as hard as others. The UV-B inhibited the growth of species in the genus *Phaeocystis* much more strongly than they affected the diatom *Chaetoceros socialis*. Here we see again the power of variation to mitigate the unforeseen and unforeseeable. The mix of plankton in the sea is extremely variable and so is its vulnerability to UV-B. This variation is helping strains of plankton and krill in the Southern Ocean to evolve and adapt in the invisible spotlight beneath the ozone hole.

What does all this mean for us? That was the instant thought of the *Origin's* readers in 1859. For answers they looked backward. They thought in terms of the past, of history, of ancestry, of lines of descent, as Darwin does in his book. All of Darwin's books are primarily about history, even where they address the present. That is the program of inquiry established by Lyell: the present is the key to the past. Lyell places that emphasis in the title of his masterwork, *Principles of Geology: Being an Attempt to Explain the Former Changes of the Earth's Surface, by Reference to Causes Now in Operation.*

Darwin's readers were shocked by what they glimpsed of our past condition in the *Origin*, and later *The Descent of Man* (also titles that look backward). But today, when we consider Darwin's process in action, we can see that our condition in the present is shocking too. We have barely begun to digest the Darwinian implications of the present moment; we have barely begun to glimpse the degree to which we are all involved in the action and reaction of evolution right now. In this sense the revolution that Darwin began in 1859 is not yet completed.

In the *Origin*, Darwin describes the record in the rocks, the virtually endless series of vanished and evaporated vistas, all those long lost ages of unexamined life, as "a history of the world imperfectly kept," of which we possess only one volume. "Of this volume, only here and there a short chapter has been preserved; and of each page, only here and there a few lines."

But we see little of the present as well. The studies of the evolution watchers are like a few scattered chapters and lines of a vast story that is now taking place all around us. Here once again Darwin's finches can serve us as symbols, heralds, and standard-bearers of events that are taking place everywhere we look and everywhere we have not yet looked. Once again the islands are showing us what is going on in our own backyards.

The changes now in progress make the whole planet and all that lives on it a single colossal demonstration of the power of Darwin's process, the kind of demonstration we think of as the domain of Darwin's islands. Many lines of life on Earth, including our own, are now living through days as charged with change as the days when the last dinosaurs died, or the first finches alighted in the Galápagos.

These changes are being driven by the rise of our own kind. As the dominant species on the planet, we are both an effect and a cause of evolution—masters and slaves of Darwin's process. We almost dread to contemplate this relationship, just as Darwin's contemporaries feared to look into the relationships his theory unfolds in the human past. But that is what the work of the evolution watchers, from the very earliest to the very latest studies, is urging us to see. The action in the present is a manifold and all-embracing evolutionary event, and the finches in their last precarious solitudes, their islands' islands, are in a peculiar position to illuminate what it means for us.

All times seem special to those who live in them. But it is neither parochial pride nor shortsighted despair to say that our time is more special than others. According to the fossil record, only five times in the past six hundred million years has there been such abrupt havoc in the biosphere. Only five times have so many twigs and branches been lopped from the tree of life at once. It happened at the end of the Ordovician period, at the end of the Devonian and the Permian, at the end of the Triassic and the Cretaceous; and now it is happening again. We are altering the terms of the struggle for existence: changing the conditions of life for every species that is coeval with our own.

Never before was such havoc caused by the expansion of a single species. Never before was the leading actor aware of the action, concerned about the consequences, conscious of guilt. For better and for worse, this may be one of the most dramatic moments to observe evolution in action since evolution began.

Chapter 20

The Metaphysical Crossbeak

How astonishing are the freaks and fancies of nature! To what purpose, we say, is a bird placed in the forest of South Cayenne, with a bill a yard long, making a noise like a puppy dog, and laying eggs in hollow trees? The toucan, to be sure, might retort—to what purpose were gentlemen in Bond Street created? . . . There is no end to such questions. So we will not enter into the metaphysics of the toucan.

—SYDNEY SMITH,
1825

Metaphysics must flourish. — CHARLES DARWIN, 1838

We stand among the other animals on this planet with a strange feeling of kinship and difference. We see so much farther and wider than they can that the eyes of Darwin's finches, in the famous engravings from his voyage, stir us to pity. They do not know where they come from, they do not know what they are, and they do not see where they are going, though birds' eyes are as sharp as our own, and though they have the advantage of their wings.

From the beginning we have tried to understand what makes us so alike and unalike: to explain so unequal a distribution of powers. In the caves of Lascaux there is a painting of a man with a beak. In the tombs of the Pharaohs there are bas-reliefs of more bird-man hybrids, the gods Osiris, Horus, and Thoth. Walking and talking in the Athenian Academy one day, Plato defined humankind as the two-legged animal without feathers. The next day, they say, Diogenes dropped by the Academy with a plucked chicken.

We are a species that has become aware of itself—which is not really, when you stop to think about it, such a very grand claim. Students of human evolution are still debating what led to this departure. Sometimes a species is carried into new territory by a journey, like Darwin's finches, borne to their volcanoes by a long flight over the sea. Sometimes it is not a journey but an invention, a novel adaptation, that opens the new world. The invention of jaws in the Ordovician period, five hundred million years ago, may have given polychaete worms the advantage over priapulid worms. Hinged jaws were a turning point for armored fish, cartilaginous fish, and bony fish in the seas of the Paleozoic, and for the whole vertebrate evolution that followed, from amphibians to reptiles, birds, and mammals.

A modest shift in the position of a few teeth, a matter of millimeters, may have led to spectacular adaptive radiations in certain lines of carnivores. In the beaks of birds, the addition of one extra hinge or flange may have led to the radiation of almost one hundred species of blackbirds, from Alaska to the tip of Argentina.

With crossbills, the elegant experiments at the University of British Columbia have shown how the first slight twist of the mandibles could have made all the difference. That mutant twist was manifest destiny. It allowed the bird to get at seeds no other bird in the forest could eat. It opened vast lands for conquest and set off a cascade of secondary adaptations. The hinge of the jaw grew specialized to allow it to move from side to side, as well as pivot up and down like normal beaks, the British ornithologist Ian Newton notes in his book *Finches*. The jaw muscles became asymmetrical to help tug the mandibles from side to side. The feet grew bigger and stronger, which helps the bird hold on to a cone while wresting the cone scales open with its beak. The crossbill also evolved new instincts, elaborate routines, and subroutines, as Newton explains:

> A bird first wrenches off a cone in its bill, carries it to some firm, horizontal branch, and clamps it between one of its feet and the perch. It then works the tip of its beak behind one of the scales, while the cone is held so that it points forwards and slightly to one side. A bird with the lower mandible deflected to the right holds the cone in its right foot, and vice-versa. . . . The lower mandible then moves sideways toward the body of the cone, so that the scale is

raised by the tip of the upper one. The seed, once released, is scooped out by the protrusible tongue.

Often the crossbill does all this work, straining and heaving, while hanging upside down in the tree.

According to present thinking, the departure of our own line began six or seven million years ago in the African savanna, when our ancestors switched from what is known in the jargon as brachiation— swinging from tree branch to tree branch—to walking on the ground. That change led to a cascade of adaptations, as with the crossbills. One of the first was the trick of rearing up and walking on our hind legs, which the evolutionist Richard Leakey calls "one of the most striking shifts in anatomy you can see in evolutionary biology."

We walked upright for millions of years before the next great evolutionary change, the expansion of the brain and the skull. This expansion, which began to take off about two million years ago, represents, like the shift onto hind legs, one of the most dramatic evolutionary changes in the fossil record. Since the days of Lucy in Hadar, the human brain has tripled in size. Meanwhile we also evolved an opposable thumb, which is the chief mechanical difference between our hands and the hands of our nearest living relatives, the orangutans, gorillas, and chimpanzees. We modified the hyoid bone, which gave us the gift of full-throated speech. There were other more cosmetic changes too, including the shortening of the muzzle, the shrinking of the jaw and the teeth, and the sculpting of the nose.

Somewhere in this sequence of adaptations (perhaps at the very start of the brain's expansion), there occurred the heightening of consciousness that we ourselves as members of the species consider distinctively human: the character that Diogenes was wordlessly pointing out to Plato. It is this character, more than the thumb, the voice, the hind-legged stance, or the human face, that we feel sets us apart from other living things on the planet. To us, a man or woman who has lost hands, legs, voice, or even face is still a human being, but a body that has lost consciousness forever has dropped from the human experience.

Some of this evolutionary action took place at the same time that Darwin's finches were radiating in the Galápagos. It also happened at about the same rate, and—despite the prejudices of human pride and power—it carried our line, in physical terms, no further from our neighbors than the finches have diverged from one another. As a sop to pride, taxonomists have placed us in a separate family from the other primates. But anatomically, chimpanzees, orangutans, gorillas, and humans are as closely related as Darwin's finches, or the two dozen species of crossbills, or many other young adaptive radiations. Chimpanzees appear to be our closest living relatives: by current estimates, ninety-nine out of every hundred genes are identical. In other words, we are as close to chimpanzees as a ground finch is to a tree finch.

"Man in his arrogance thinks himself a great work, worthy the interposition of a deity," Darwin scribbled in one of his first secret notebooks, when he had become convinced of the fact of evolution. Darwin felt it was "more humble & I believe true to consider him created from animals." Our gift of consciousness is a mystery, one of the greatest remaining mysteries in biology—but it is no more of a miracle than a beak, a feather, or a wing, and it is made by the modeling and molding of the same living clay, through the same process, Darwin's. Why should we assume that consciousness is unique to our kind in anything but degree? "It is our arrogance," Darwin wrote in his notebook, "it is our admiration of ourselves."

Someday neurobiologists hope to close in on the origin of consciousness in the brain. They will find some twist in the neural networks of the frontal lobes or the cerebral cortex that leads, as it grows, to a kind of infinite recursion, rather the way mirrors tilted at the proper angle will begin to reflect each other. The discovery of the physical basis of this secret may still be far away, or it may be nearer than we think. But as with the crossbill, it will probably turn out to be a twist in equipment that we otherwise share with many other species: a twist that heightens the trick of recursion and enables us to do tricks with our awareness of the world that other animals cannot do, to pick things up that no other species can handle.

Perhaps when biologists have sequenced the whole of the human genome, and deciphered many of its messages, some slight difference between our genes and the chimps' may throw more light on this mystery and help us understand the cerebral kink that has made us the metaphysical crossbeak.

One of the gifts of our heightened consciousness is the ability to make new tools. We can build our own adaptations and evolve, in effect, within our own lifetimes. We once imagined this gift, like consciousness, to be uniquely human; Benjamin Franklin called our kind *Homo faber*, Man the Toolmaker. But Bonobo chimps make and use tools too, and so does the woodpecker finch of the Galápagos, which picks and chooses cactus spines to augment the beak it is born with. Here again we differ from other species only in degree. The difference began to widen thirty or forty thousand years ago, long after our brains and skulls had expanded to their present volume. Suddenly, in southern France and northern Spain, we began carving bone and knapping flints so cleverly that the tool kit of even a single hunter surpassed the beaks of all Darwin's finches: bone awls sharper than a sharp beak's, stone chisels bigger than a big beak's. By now, of course, in our own moment in the evolutionary play, the competitive advantages of our toolmaking are on display all over the planet, even in the Galápagos, where members of our species propel themselves on webbed feet beneath the parrot fish; fly higher and faster than the boobies and frigatebirds; skim the sea faster than the dolphins; and go to sleep each night far out on the water, where the lights of yachts now shine in the eyes of birds along the cliffs like newly risen stars.

With our heightened consciousness we have been able to carve out more adaptive niches more rapidly than any other species on the planet, radiating out of Africa to every continent and both poles. We can do this because our suite of adaptations—the big conscious brain, full-throated speech, and opposable thumbs—allows us to invent new ways of living, not only new tools but new foods, clothes, and shelters, and to pass them on to others with unprecedented speed. Once again the difference is one of degree, as the biologist John Tyler Bonner argues in his recent books *The Evolution of Culture in Animals* and *Life Cycles*. In Great Britain, blue tits have learned to peck through the aluminum caps of milk bottles on doorsteps and get the cream. People actually saw the trick spread from house to house and block to block as blue tits across the country learned from watching one another. For a while the blue tits had an easier life than Darwin's finches wrestling with *Tribulus*, or crossbills with green pinecones. But soon the milk companies, like the plants, will start putting out their treasures with tougher caps, if they have not done so already.

A famous young macaque monkey named Imo, on an island off the coast of Japan, learned to wash sweet potatoes in the sea before eating them. She also learned to separate wheat kernels from sand by cupping them in her hands and dipping them in the water. The other macaques on the island picked up both these tricks by copying Imo.

Recently investigators in Italy allowed an octopus in a tank to watch through a window as another, trained octopus performed a trick, choosing a red ball or a white ball. If the trained octopus chose the correct ball, it found a small piece of fish behind it; if it chose the wrong ball, it got an electric shock. The experimenters videotaped the untrained octopus. It was watching closely, following the action with its head and eyes, time after time. Afterward, when it was given the same choice, more often than not, it chose the correct ball.

In a social species like ours the benefits of learning from one another are so great that they may have helped to drive the modification of the hyoid bone, toward better and better vocalization, leading at last toward speech; and the expansion of the brain may have been driven in part by the increasing benefits of and demands of language. Language became a tool that allowed us to teach not only each other but even ourselves, after we found a new use for the opposable thumb and fingers. "When you write," says Annie Dillard, in *The Writing Life*, "you lay out a line of words. The line of words is a miner's pick, a woodcarver's gouge, a surgeon's probe. You wield it, and it digs a path you follow. Soon you find yourself deep in new territory."

The ability to learn new tricks from one another is called cultural evolution, and clearly it is not unique to our kind. Perhaps the most haunting glimpse of cultural evolution in another species comes, once again, from Darwin's finches—but from a species that Darwin never saw.

There is a solitary finch species that lives outside the Galápagos. Its home is on the speck of land that lies nearest the archipelago, 630 kilometers (almost 400 miles) northeast, a little place called Cocos Island.

Like the Galápagos, Cocos is volcanic. The island has virtually no coast: it is walled by steep cliffs as much as 180 meters high, almost all the way around. No human beings have ever tried to raise families there, and it is unlikely that anyone will ever try, because the place is almost as offputting in its way as Daphne Major. Unlike Daphne, however, it is drenched with rain almost every day. The island receives an astonishing 7 or 8 meters of rain a year, and lush rain forest grows from the summit to the very edge of the cliffs.

In some way Cocos finches are more diverse than Galápagos finches. Some eat bugs, others crustaceans, still others nectar, fruits, seeds. There are specialists that eat mostly on the ground, others in bushes, others in tall trees. Normally to assemble a list of birds with such a wide range of specialties one would have to collect not only many species, but many genera, or even many families, groups of genera. Yet the finches of Cocos Island are a single species.

These finches cannot diverge and ramify on Cocos the way their siblings have in the Galápagos archipelago. On Cocos Island they can't get away from one another. The place is too small, and the nearest land is too far away (Cocos is about 500 kilometers, more than 300 miles, from the coast of Costa Rica). There is no chance for geographic isolation. Insects and snails can radiate in such a place, because for them even a tiny island is big enough to present, in effect, an archipelago of isolated habitats, but birds do not radiate on single isolated islands.

That, of course, is why our own kind is not radiating into new species. For us the whole planet is almost as small now as Cocos is for the Cocos finch.

Rosemary and Peter Grant have not studied the Cocos finches. The most careful work on them to date has been conducted by another married couple of evolutionists, Tracey Werner and Tom Sherry. In the 1980s they spent a total of four seasons watching the finches, concentrating on a single big thicket of hibiscus. There they banded, watched, and measured about one hundred finches, in the style of the Grants (in fact, Nicola Grant was one of their field assistants). They watched their hundred finches eat, season after season, recording and analyzing a grand total of 26,770 feeding attempts.

The bodies of the finches on Cocos Island are far more uniform in size and shape than in the Galápagos, and they all carry the same pointed, slender, all-purpose beak. Their specialties have nothing to do with their sex or age, or with minute variations in the size and shape of their beaks. Nor is a Cocos finch's food choice tied to the seasons, the time of day, or some particular spot in the hibiscus thicket. Werner and Sherry saw the finches working away at their chosen trades indefatigably at all times and places; they often saw half a dozen finches each going after food in his or her own way at the same time in the same bush, like blacksmiths, bakers, printers, and tailors in the shops around a village green. Some gleaned bugs from branches, some probed the branches, some gleaned from living leaves, some from dead, curled leaf clusters, and some sipped nectar from flowers. Whatever their métier, they kept at it hour after hour, day after day.

Most of these trades required skilled labor. For instance, Werner says, Cocos finches that probed branches for bugs had to pry, strip, and twist bark off the branch to get at the insects below, as tree finches do in the Galápagos. She says many of the finches' trades were not only skilled but highly specialized: some finches in the hibiscus thicket spent most of their time picking out tiny moth larvae from between leaf layers in morning glories.

Werner and Sherry suspect that the finches were learning all these specialized trades in the way that human beings do: from their elders. The finch watchers often saw a young bird hopping along a short way after a grown one, watching and copying. The juvenile would observe, hop to exactly the place the adult had just left, and do what the adult had just done at that spot.

Werner and Sherry also saw the juveniles hopping along after warblers and sandpipers, watching and copying them. The young fledglings also flocked together, in groups of anywhere from two to thirty, like teenagers at a shopping mall, and they watched and copied one another too.

Evolutionists know that lions, apes, elephants, and other social mammals learn from their elders, but they tend to think of birds as inheriting their skills simply through instinct. In the staggeringly rich rain forests, however, as the evolutionist Jared Diamond has observed, there may be an adaptive advantage for birds that can learn from their elders. "An insectivorous bird is faced with tens or hundreds of thousands of local insect species," Diamond notes, "including more than a thousand beetle species in one tree species alone." Even trained human beings—rain-forest shamans and entomologists with Ph.D.s in rain-forest ecology—have to spend decades, entire lives, learning to recognize a small fraction of this diversity. Perhaps finches in the Cocos rain forest have been selected for their learning ability, just as human beings were selected for it at the dawn of our species.

Compared with other birds, the Cocos finches struck the finch watchers as awkward, even clumsy, feeders. They often saw the birds lose their grip while looking for bugs, and fall out of the trees. Sherry once watched a juvenile trying to peel bark from a branch. The finch fell, hovered, perched, peeled, fell, hovered, perched, peeled—over and over, until he finally extracted a prize, a one-inch-long centipede. Another time, Sherry saw a finch trying to catch a spider on a branch. "The bird lunged for the spider and missed," Werner writes in her thesis on the Cocos finches. "The spider dropped down, hung from its thread. The bird scrambled down the trunk of the tree." Each time it came alongside the spider the finch took another stab and missed again. The spider kept spinning out thread, and the bird kept hopping alongside it trying to keep up: leaping, lunging, missing, all the way down the tree to the ground. There the spider made a mad dash—and another finch got it.

A finch watcher once saw a Cocos finch give the same bumbling kind of chase on a liana. "It seemed odd," he wrote afterward. If the bird had used its wings it probably could have outmaneuvered the spider very quickly, he thought. Yet it chose instead to sideslip slowly and awkwardly down the vine.

This sounds like the sort of comedy that other animals might enjoy every day if they watched us doing what we do. Compared with fish we are bad at swimming, compared with birds we are stiff at flying, compared with cheetahs we are ludicrous at running, compared with ants we are hellacious at cooperating. Yet we are the most successful species of our time. We have overrun and overturned the territories of all these other animals because taken as a whole, by learning from the generation before us, we can do a fair job at all of their skills at once. As the evolutionist Ernst Mayr has written, we have "specialized in despecialization."

Our position on the planet is the same as the Cocos finches in their private rain forest. The range of foraging opportunities that lies open to us vastly exceeds our ability as individuals to take advantage of them all. Like the finches we have evolved an extraordinary ability to learn, so that as a species, collectively, we can take advantage of these myriad niches, and we keep finding more and more trades. We fill more different ecological niches than any other animal.

This is what allows us to carry on the epic learning game we call science. Science formalizes our special kind of collective memory, or species memory, in which each generation builds on what has been learned by those that came before, following in each other's footsteps, standing on each other's shoulders. Each generation values what it can learn from the one before, and prizes the discoveries it will pass on to the next, so that we see farther and farther, climbing an infinite mountain.

"I FIND IN [the] animal kingdom," Darwin scribbled to his botanist friend Hooker, "that the proposition that any part or organ developed normally (ie not monstrosity) in a species in any *high* or *unusual* degree, compared with the same part or organ in allied species, tends to be *highly variable*. I cannot doubt this, from my mass of collected facts.—To give instance, the Cross-Bill is very abnormal in structure of

Bill compared with other allied Fringillidae, & the Beak is *eminently variable.*"

Hooker wrote Darwin that he did not see this phenomenon in plants. Darwin shot back, "I daresay the absence of Bot. facts may in part be accounted for by the difficulty of measuring slight variations. Indeed after writing this occurred to me; for I have *Crucianella stylosa* coming into flower & the pistil ought to be very variable in length, & thinking of this I at once felt how could one judge whether it was variable in any high degree. How different, for instance, from Beak of Bird!—"

Darwin develops this theme in the *Origin* in a chapter titled "Laws of Variation," under the heading, "*A Part Developed in Any Species in an Extraordinary Degree or Manner, in Comparison with the Same Part in Allied Species, Tends to Be Highly Variable.*" Here Darwin cites the beaks of domestic pigeons, when the beak is part of what distinguishes the breed.

This pattern holds across nature, and of course the beaks of Darwin's finches have turned out to be a particularly dramatic case. From the moment they hatch, these species differ more from one another in their beaks than in any other character the Grants have measured. No other character in Darwin's finches is as variable as the beak.

In our own species, it is the human brain that has diverged most dramatically from those of our nearest relatives on the tree of life. True to Darwin's law the human brain is exceedingly variable in its measurements, like the beaks of Darwin's finches, pigeons, and crossbills. The volume of the human skull, its cranial capacity, is more variable than the depth of *fortis* beaks on Daphne Major.

The mind is our beak, and the human mind is even more variable than the brain. Darwin, for instance, was extraordinarily gifted at observing, collecting, theorizing, and also at finding mentors. But he was a poor mathematician. He wrote to a friend at school who had not answered his letters, "I suppose you are two fathoms deep in mathematics, and if you are, then God help you, for so am I, only with this difference, I stick fast in the mud at the bottom and there I shall remain." Which turned out to be true. (That is one reason Darwin's theories are more accessible than Newton's. He expressed even his deepest thoughts in a language that anyone can read.)

Not everyone in a village or neighborhood can master every trade, and no one man or woman can master more than a few, but if people specialize, then collectively the villagers have a hundred trades. Their infinite variety of minds and talents helps them to radiate into all these crafts and specialties.

"I believe that this psychological polymorphism has been a major reason for the success of the human species," says J. B. S. Haldane. Variation is the secret of our adaptive radiation. There is infinite value for the species in our infinite variety. As among Darwin's finches, it is the tool by which we get more of the food available to us than we ever could if each of us tried to do it all, if each human being tried to be a generalist. Our minds and talents are variable for the same evolutionary reason as finches' beaks are variable in the Galápagos: jack-of-all-trades, master of none. And what drives this radiation within our species is a process like character divergence. Though we may not think of it as Darwinian, we all feel its pressure, wanting and needing to do what we are made for—seeking the task for which we are most fit.

Emerson writes, "Each man has his own vocation. The talent is the call. There is one direction in which all space is open to him. He has faculties silently inviting him thither to endless exertion." William Blake: "How ridiculous it would be to see the Sheep Endeavoring to walk like the Dog, or the Ox striving to trot like the Horse; just as Ridiculous it is to see One Man Striving to Imitate Another. Man varies from Man more than Animal from Animal of different Species." Aeschylus: "Character is destiny."

Those whose lives are comfortable believe themselves more or less insulated from the pressures of natural selection, because they have the leisure and the freedom to make so many choices in their lives. But they are ruled by selection as much as any other creatures on the planet, for they too are using their own individual variations to lessen the pressure of natural selection. Everywhere very young human beings start out acting more or less alike, just as fledglings on Daphne start out using their diverse beaks in more or less the same experimental ways. As we get a little older we enter a phase of wild experimentation, as finches also do on Daphne. As we get older yet we narrow our efforts, again like the finches. In every country, within the limits of our choices and opportunities, we tend to seek trades in which we have learned by experience that we are unlikely to lose, to be killed or driven out by the competition, trades in which our weaknesses will do us minimal harm. We try to find the work for which our beak best fits us—although what we find in the end is rarely perfect, as Darwin's contemporary, the witty parson Sydney Smith, observes,

If you choose to represent the various parts in life by holes upon a table, of different shapes—some circular, some triangular, some square, some oblong—and the persons acting these parts by bits of wood of similar shapes, we shall generally find that the triangular person has got into the square hole, the oblong into the triangular, and a square person has squeezed himself into the round hole. The officer and the office, the doer and the thing done, seldom fit so exactly that we can say they were almost made for each other.

From which the phrase "A square peg in a round hole." And as Mark Twain adds, "A round man cannot be expected to fit in a square hole right away. He must have time to modify his shape."

IN THIS WAY we have become the evolving animal. We are now evolving rapidly ourselves, and we are driving evolution everywhere around us. We have learned how to make Darwin's process run faster for us than it does for any other species on the planet—except perhaps the bacteria, with their flying rings of plasmids and ten-minute generation times. The tragedy of our success is what we are doing to the rest of creation, which evolves more slowly.

Our own tenure has been brief, and on average the term of a species is brief—a few million years. A species that can survive only by causing upheaval around it is in constant danger of extinction, like a tribe that lives for battle. At the moment the whole planet is like a closed pinecone that we alone with our twisted beaks have contrived to open, so that there are more of our kind now than of any other bird in the forest. Yet the rapid accumulation of change is not always progress, and forward motion is not always an advance.

When they visit cactus flowers, cactus finches on Daphne Major sometimes snip the stigma, which is the top of the hollow tube that pokes out like a tall straight straw from the center of each blossom. When the stigma is cut, the flower is sterilized. The male sex cells in the pollen cannot reach the female sex cells in the flower. The cactus flower withers without bearing fruit.

These finches are, of course, completely dependent on the cactus. Without cactus pollen, cactus nectar, cactus fruits, and cactus seeds, they would starve. The birds' fates are so closely bound up in the fates of the cactus that when there is more cactus on Daphne Major, there are more cactus finches on Daphne Major; when there is less cactus, there are fewer finches.

Stigma snipping.
Drawing by Thalia Grant

On Daphne one December near the start of the cactus-flowering season, the Grants peeked into more than two thousand cactus flowers. Almost half the flowers had lost their stigmas. On Genovesa the next month the Grants examined more than a hundred cactus blossoms, and four out of five of them had been violated. In some years the cactus finches have destroyed almost every cactus flower, and in those years the cactus on their island produced virtually no fruits or seeds. It is hard to imagine a simpler, neater, faster way for a species of Darwin's finches to drive itself toward extinction.

To find out what was going on, the Grants took turns, rotating every two hours all day long, keeping watch over seventeen cactus flowers. They noted when each flower opened, which finch fed at each flower, and what each finch did. Cactus flowers usually open in the morning between 9 and 11. When a cactus finch lands beside an open flower, it holds the stigma to one side with its foot so that it can nibble on the pollen that is cupped in the base of the flower. But sometimes a cactus finch will visit a cactus bud very early in the morning, an hour or two before opening time, and pull apart the folded petals to get in there before anybody else. When the flower is pried half-open, the stigma is liable to poke the finch in the eye. On Genovesa, in the year of their stigma study, the average stigma stuck out of the cactus flower about 25 millimeters; the average distance from the tip of a cactus finch beak to its eyeball was only 21 millimeters. So, when it forced open the flower bud, the finch would sometimes snip the stigma with its beak and flick it away. Not all of the cactus finches on the island were doing this, the Grants discovered: only about a dozen.

A finch that snips stigmas is like a farmer who eats seed corn. The bird steals from its future and the future of its line. By sterilizing flowers, the cactus finches on Daphne Major cut each year's harvest in half. And all the stigma snippers get out of it is a little pollen the other birds can't get, and a bit of nectar the other birds can't reach—a sweet treat in the early morning.

Darwin's process cannot stop this dirty dozen, and in fact the process favors them, because the stigma snippers pay no special price for their stolen sweets. They are no more likely to die in the dry season than the birds that spare the stigmas. In dry seasons, cactus finches forage on one another's cactus anyway; they don't stay within the borders of their own territories when they are hungry. So the dozen flower violators on the island are trampling on a commons, not on their own private gardens. A finch that spares the flowers on its territory cannot guarantee itself a good meal later in the year when times are hard. In fact, as the Grants point out, a bird that takes good care of all of the stigmas on its property may even encourage trespassers later.

A cactus finch. From Charles Darwin, *The Zoology of the Voyage of H.M.S. Beagle.*
The Smithsonian Institution

Natural selection turns upon the profit of the individual. What is good for the individual is usually good for the flock. But when the needs of the individual clash with the needs of the flock, it is the individual that triumphs, even if this private success leads to the downfall of the flock. If the terrible drought of 1977 had been followed by a second year just as dry, all of the cactus finches on Daphne would have been in jeopardy because of the stigmas that the dozen birds had severed. Those birds might have made the difference between survival and extinction on Daphne Major.

"The great God that formed all things Both rewardeth the fool, and rewardeth transgressors." Each year the habits of the dirty dozen on this island handicap the flock and increase the odds that the whole flock will die. Cactus finches did go extinct earlier in this century on the uninhabited island of Pinzón, whose name is Spanish for *finch*. The cactus finches of Pinzón may have cheated themselves off the face of the earth.

DARWIN HIMSELF was (at least sometimes) an optimist. In the *Descent of Man* he wrote,

> The Simiadae then branched off into two great stems, the New World Monkeys and Old World Monkeys; and from the latter at a remote period, Man, the wonder and the glory of the universe, proceeded.

He also wrote,

> Progress has been much more general than retrogression.

Maybe Darwin was right about progress, and his descendants the G.O.D. specialists are right. Maybe the Generation of Diversity will outrace the Generation of Destruction. In the long run we may yet prove to be more children of light than children of darkness, despite the tinny hubris of the acronym G.O.D.

On the surface of the earth, on the surface of the backs of our hands, there are ropy loops of DNA in every cell, a changing galaxy of atoms in every strand of DNA, and views that bring a sense of origins as ultimate as the Big Bang. There is nothing more absorbing than to look around at trees, bare stubbly fields, a big black turkey vulture wobbling overhead, from a winding strip of asphalt, at 60 miles per hour, and reflect that all this living scene is in motion too, in ways we can now begin to see. The branching of the branches of the tree of life—including our own branch—all this branching is going on right now, everywhere, although it is hidden from our view like stars at noon.

We have watched the animals around us from the beginning, each generation learning from the generation before, so that each sees more than the one before—as in these islands, where the first voyager to write about the birds said they showed neither novelty nor beauty, and where Darwin himself declared it impossible to watch the species of ground finches and tell them apart. Now these creatures' evolution is the best known and the best watched on the whole of the planet, and they teach us what is going on in all the rest.

This is what we do best. We add to what was learned before, raising the old questions again and again, lifting them if we can toward higher and higher ground, and ourselves with them. Why are there so many kinds of animals, and why are we among them? Probably we have been asking these questions ever since we lived in caves—when it was still the common experience of our kind to stand alone on a cliff's rim and survey a wilderness of animals, feeling kinship in difference, difference in kinship, our eyes watchful in their large orbits, our arms outspread. The questions were born then, as we stared down, or looked up at the sky, turning to follow the turning flocks, revolving our heads on our featherless shoulders: high above the plains of our first hours, winged only with the questions.

Epilogue

God and the Galápagos

Nature is the art of God.

— SIR THOMAS BROWNE,
Religio Medici,
1642

Have we any right to assume that the Creator works by intellectual powers like those of man?

— CHARLES DARWIN,
Origin of Species

It is March 1993 on Daphne Major. The craterlet is filling up with water. Boobies are swimming around in it like ducks. The paths are covered with grass and flowers—more grass and flowers than Rosemary and Peter have ever seen on the island. If it were not for the cactus, the Grants would almost think the Galápagos had gone adrift, true to their old name, Las Encantadas, the Enchanted Islands, and wandered into the Temperate Zone.

Only a few years ago the Grants could walk all the way around the rim of the main crater and not see a single finch. They couldn't do that now. They are tripping over them now. The birds are incredibly bold too. "Do you think they've got tame, or are there just more of them?" Rosemary asks Peter. All day, Darwin's finches fly up from the lava and perch on their shoulders. They come zipping in and land on Peter's head.

Last year was a Niño. The year before that was almost a Niño, but then it fizzled. This year is even wetter than last year. "A string of three extremely wet years," says Peter. "We have never seen anything like this. It is unprecedented in our experience."

"I don't know what they are saying anywhere else," says Rosemary, "but on Daphne, this is a Niño!"

That is one of their catchwords this season, as the rain falls and falls: "I don't care what they're saying. It's a Niño!"

This is their twenty-first year on the islands. When Peter and Rosemary first came here they were still in their thirties; now they are almost in their sixties. They had been dreading the return of El Niño: they wondered if they would be able to band all those birds. Now they are proving to themselves that they can still do it, alone on the island, with no assistants, for the second year in a row. But Peter's beard has lost the last trace of auburn. The top of his head presents a somewhat larger landing zone. If they get back to Princeton this year, his friends will be shocked how thin he looks.

When rocks rattle and roll underfoot, they reverberate. Otherwise there is no sound but the wind in their ears and bird cries nearby and far, far down in the crater. "We have more birds breeding on the island this year than any other," says Peter. "Except possibly 1984."

The first rogue that Rosemary trapped on the north rim, two Januarys ago, is dead now. The body was never found. But Rosemary's second catch that morning, 5608 (the Princeton bird, orange-over-black, born in the great Niño of 1983), is ten years old. Not only that, this year he was one of the first to breed. "5608 is in good health," Peter says. "Got off to a flying start. Also, do you remember 2666? He turned up again this year. He's the oldest *fortis* we have on record. That pushes the maximum life-span of *fortis* to fifteen years, which is really quite remarkable."

"2666 *bred* this year," says Rosemary.

"Hmmm," says Peter, in comfortable agreement. "Still going strong." He is surprised the old black cocks like 5608 and 2666 can find mates. Some of these birds wear their crowns half-bald rolling rocks around in the droughts, and then the bald spots grow back with brown patches. "I wouldn't, if I were a female, mate with one of those," he says.

"And they get clobbered by younger males," Rosemary says.

"But then, look at me," says Peter.

THIS YEAR A poll will show that nearly half the citizens of the United States do not believe in evolution. Instead they believe that life

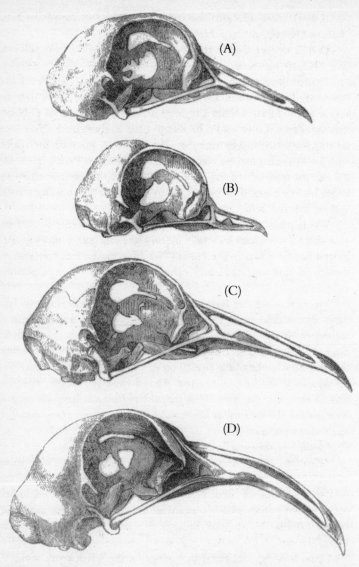

The power of selection, as seen in the skulls of pigeons. The common rock pigeon (A) has produced forms as various as the short-faced tumbler (B), the English carrier (C), and the Bagadotten carrier (D). From Charles Darwin, *The Variation of Animals and Plants under Domestication.*
The Smithsonian Institution

was created by God in something like its present form, within the past ten thousand years.

"People talk about Creationism," says Dolph Schluter. "We can actually see Creation at work. We might ask the Creationists to demonstrate similar principles at work."

"They have the appearance of closed minds," Peter Grant says. "I don't often meet fundamentalists, and I don't go out of my way to challenge them. I think they are closed in that respect."

John Endler, the guppy watcher, does not like talking with Creationists either. "I avoid it," he says. "It's really a waste of time. Not long ago on an airplane I talked for an hour with someone about what I do, and never once mentioned the word *evolution*. It's very easy to do, you know. Darwin himself doesn't use the word *evolution* in the whole of the *Origin*. You just talk about what happens, and how you can study what happens: changes over many generations. It might be interesting to try to write a book that way now: don't use the word *evolution* until the very last page.

"Anyway, the whole time on the plane, my fellow passenger was growing more and more excited. 'What a neat idea! What a neat idea!' Finally, as the plane was landing, I told him this neat idea is called evolution. He turned purple."

"I've done exactly the same thing—and never let on it was evolution—and got exactly the same response," says Rosemary. "I described our work on Daphne to a Jehovah's Witness. And he followed along, and said, 'Oh, how fascinating.'"

"Asked intelligent questions," says Peter.

"And I never plucked up enough courage to say, 'Well, you know, what all this means. . . .'"

Darwin, of course, was surrounded by Creationists too, including some of his greatest friends and mentors. Lyell, the geologist, after visiting Darwin's pigeon coop, argued that however animals might or might not have been created, surely human beings were designed by the intervening hand of God. But Darwin asked if Lyell could believe that "the shape of my nose was designed." If Lyell did think so, Darwin said, "I have nothing more to say." If not, seeing what breeders had accomplished by selecting slight variations in the nasal bones of pigeons, why should our own beaks have needed anything more? Selection can make a beak; selection can make a nose.

"If anything is designed, certainly man must be," Darwin wrote to another devout friend, the botanist Asa Gray. "One's 'inner consciousness' (although a false guide) tells one so; yet I cannot admit that man's rudimentary mammae, bladder drained as if he went on all four legs, and pug-nose were designed. If I was to say I believed this, I should believe in the same incredible manner as the orthodox believe the Trinity in Unity. You say that you are in a haze; I am in thick mud; the orthodox would say in fetid, abominable mud; yet I cannot keep out of the question, My dear Gray, I have written a deal of nonsense."

Once it seemed logical to believe that God shepherds the planets around and around the sun. Regular orbits were said to be proof of the existence of God, a celestial argument from design. Astronomers imagined an invisible hand in constant attendance, pushing and rolling each world through the sky. This vision no longer seemed compelling after Galileo and Newton discovered the celestial laws of motion (consider, for instance, inertia).

Darwin discovered laws of terrestrial motion as simple and universal as the physicists'. For Darwin it was no longer necessary to assume that God's hand had shepherded each line of life individually into being and molded it like clay. Paley's argument from design collapsed. But beyond this, Darwin admitted ignorance.

"I feel most deeply that the whole subject is too profound for the human intellect," Darwin wrote Asa Gray. "A dog might as well speculate on the mind of Newton. Let each man hope and believe what he can."

TREVOR PRICE IS watching warblers in Siberia. ("Siberia is really opening up," he says.) Lisle Gibbs is studying the DNA of cuckoos and cowbirds; Peter Boag the DNA of Darwin's finches; and Laurene Ratcliffe the songs of sparrows, chickadees, and red-winged blackbirds.

For Dolph Schluter several ages have passed since he was watching finches on Pinta, listening with headphones to the Clash. Now he is watching fish in tanks and listening to *La Traviata*. He spends whole days with the sticklebacks in his lab and out on Paxton Lake. When he filled the new evolution ponds on campus, not long ago, he popped the cork of a bottle of Veuve Clicquot. "We plan to watch for twenty years," he says, "or whichever senesces first, me or the experimental ponds."

Ars longa, vita brevis. A life turns out to be a very short time. Dolph sometimes thinks of his hero, David Lack. "This is just a rumor that he said this," Dolph says. "I heard it second- or third-hand—I've never

seen it written up—but it was concerning the Great Tits of Oxford."
Someone asks Lack a question about the tits in Wytham Woods.
" 'Well, I can't answer that question,' Lack says, 'because I have only
seventeen years' data.'

"That doesn't surprise me any more. I thought it was hilarious
when I started grad school, but it's pretty obvious to me now."

On Daphne, Peter and Rosemary are now preoccupied with birds
in the 18,700s. "What was 18,717 doing yesterday?" Soon they will be
in the 18,800s. By the end of this spring, the Grants will have seen
two dozen generations of finches hatch and die on the sides of the old
bowl, their crater island, which the younger finch watchers used to
call "the biggest ashtray in the world." The Grants will be able to measure the powerful winds of selection that have blown across those two
dozen generations, shaping the wings, legs, and beaks of the birds. But
what are two dozen generations in the Galápagos? Daphne Major, a
young rock as rocks go, is almost a million years old. It is the cinerar-
ium of innumerable generations. A thousand ages in this place are like
an evening gone. And the trail of the generations extends out of sight,
both behind and ahead, like a line of flying birds that crosses the sky
from horizon to horizon, or like the zephyrs that the Grants can see
from the island's rim on the calmest days, brushing the face of the sea.

From every line of living things, not only the birds but also every
plant and animal, the same line of generations extends from horizon to
horizon, "without ever stopping," as in the pitying words of the Buddhist scripture: "Over and over again they are born, they age, die, pass
on to a new life, and are reborn!"

Our own journey has not been tried before. No other creature has
traveled so far along this line. No one knows what paths still stand
open, if we manage to travel a little farther yet. The long columns of
code in the Grants' waterproof notebooks look random, formless, chaotic, like the lava. Only when the Grants take a step back from the
numbers, only months or years from now, will the latest patterns
emerge, the special providence in the fall of this spring's sparrows.

The original meaning of the word *evolution*—the unrolling of a
scroll—suggested a metamorphosis, as of moths or beetles or butterflies. But the insects' metamorphosis has a conclusion, a finished adult
form. The Darwinian view of evolution shows that the unrolling scroll
is always being written, inscribed as it unrolls. The letters are composed by the hand of the moment, by the circumstances of the day itself. We are not completed as we stand, this is not our final stage.

There can be no finished form for us or for anything else alive, any thing that travels from generation to generation. The Book of Life is still being written. The end of the story is not predestined. Our evolution continues. Like Belshazzar, when we look around us now we can almost see "the part of the hand that writes."

WHEN THE ANCESTORS of Darwin's finches landed here, these islands were new. They may have been the first living things to try the strange fruits of the islands and pick up the seeds from the lava; the first to perch in the half-naked bushes; the first to sleep, beak under wing, in the cactus trees.

A woodpecker finch becomes possible only on an island without a woodpecker, a warbler finch only without a warbler. A flower-browsing finch becomes possible where there are no bees and hummingbirds—and on islands where bees have now invaded, many of Darwin's finches have given back the flowers. Many paths lay open when the finches first arrived, and the smallest flights and trials of their descendants were rewarded. That is why they have traveled in more directions than any other creatures on the islands, that is why they have evolved farther and faster than any other creatures: because they got here early.

Our own line is now radiating farther, faster, and in more directions than any other single species in the history of the planet—and for a similar reason. We are the first creatures to arrive in the strange territory we now occupy. We stumbled into our new niche before any other creatures on the planet. We discovered it.

"There came a flocke of birds into Cornwall, about Harvest season, in bignesse not much exceeding a sparrow, which made a foule spoyle of the apples," wrote Richard Carew in 1602. "Their bills were thwarted crosswise at the end, and with these they would cut an apple in two at one snap, eating onely the kernels." Like a sudden flock of birds, our wings fill the land. We hold all the fruits of the planet in our beaks.

Chimpanzees in the last of the African rain forests are teaching their children how to crack nuts with stones. They are taking the kinds of halting, modest steps that might, if favored, if selected for many generations, carry them on a path toward something like our niche. Watching them, human observers notice striking variations in intelligence from individual to individual. Where there is genetic variation, there is room for selection and evolution. Probably in every line of primates there are a few, like Imo on her island, who could lead their kind in our direction. But at the moment of course that path is blocked. The thinking niche is taken, at least for the moment.

Possession, as we say, is nine-tenths of the law. This rule applies to all life on earth. Before life began, our planet was like the Galápagos Islands before the arrival of the first seeds: newly cooled, newly hardened, and vacant. The first molecules that could make copies of themselves could grow at their own pace. A molecule that duplicated itself erratically, budding off carbon copies with individual variations, could grow faster and faster and sweep the seas, infiltrating myriad cupped coasts and cracked ocean beds, taking off in new directions everywhere. All paths lay open to those changelings; the path of paths lay open. The molecule that began the journey had the form of a helix.

Biochemists can now make primitive self-replicating molecules in computer models and synthesize them in the laboratory. Some of the new designs are based on the helix, and some are hopeful molecules as fanciful as Darwin's pouters, laughers, and fantails. This is a new kind of artificial selection, selection for life itself.

In this way, even now, not only in test tubes but all over the planet, matter still makes tentative steps toward life. In the shallows of Darwin Bay, and in volcanic vents at the bottom of the sea, the soup is still warm. There are more inanimate molecules in the seas of this planet than there are stars in the universe. Here and there, now and then, a few of them begin to make the kinds of connections that— again, if favored, if selected generation after generation—would bring them to life. Matter may take these first steps daily and hourly in the hot springs on the sea floor beneath the Galápagos, in the deep, drowned rifts that geologists call Darwin's Faults.

In the laboratories, the trial soups are kept hermetically sealed, or each experiment would be cut short before it got interesting because the new molecules in the soup would be scavenged by bacteria. The waiting Pyrex ponds are sterile as the seas and shorelines of this planet before life began. But in the ocean, of course, as fast as molecules make their first gestures toward life, they are devoured. Creation in the sea has never stopped, but the niche of life is taken.

That is why we look around us now at other planets and other suns. Are we the first, or are we only the first on this island? Are we the only ones to have evolved the twist of replication with variation—

and beyond that, the twist of heightened consciousness, the crossed beak in which we feel we can grasp whole worlds?

This is the not-so-secret mainspring of our fascination with aliens and extraterrestrial intelligence, in the century of our first wings. What we have here is a species calculating its chances and its future quite consciously and coolly: a whole species before a big move. We are perched on one rock, gauging our chances of reaching another. Our juvenile space fantasies are a species' dreams of invasion, expansion, and domination—and they may come to pass. Someday we may sail among new enchanted archipelagoes with *Tribulus* seeds clinging to our boots.

At the same time we fantasize endlessly about first encounters here on earth. This is just what Darwin's finches would do if finches could fantasize. We know now that at any moment we ourselves may be surprised, like the finches of Santa Cruz among their anis, goats, cats, and fire ants; or like the old Easter Islanders, who once believed their solitude was greater than it was, that they were the only human beings on this earth.

This year in the population explosion on Daphne Major, the Grants in their camp on the east rim of Daphne Major see more and more young finches winging out to sea, sometimes one by one, sometimes in small flocks, and making for the nearest moon-colored rock on the horizon, Daphne Minor. Some head for Santa Cruz, flying out and then turning around and coming back again. A few go on and on and never come back. That is our position now on the archipelago of the continents, crowded, restless and anadromous, looking up at the stars.

DARWIN'S FINCHES MAY have made more journeys than we ever imagined. Not long ago a team of geologists surveyed the sea floor around the Galápagos. Away to the east, in the direction of the South American continent, they report in *Nature*, "we dredged abundant well-rounded basalt cobbles from a small seamount with a terraced summit." They found the summits of a few other volcanoes at the bottom of the sea and dredged cobbles from them too. Though the cobbles are stained greenish-brown from lying in the depths, they look much like the worn lava rubble that lies at the feet of all the cliffs in the Galápagos. Lava does not grow in that shape at the bottom of the sea; it is worn into cobbles by the rolling of waves, like pebbles on a beach. The cobbles suggest that these drowned islands once stood above the sea. Their ages, according to the geologists' report, range from five million to nine million years. The oldest peak lies submerged at 85° longitude, which is halfway from the Galápagos to the continent. When it stood above the waves, it may have been about the size of the island of Pinzón. It rises steeply to a generally flat summit, the geologists say, with a small peak toward the northwest edge. They see hints of a number of wave-cut terraces at different heights, "and a residual spire." In profile, except for the spire, it is not unlike the shape of Daphne Minor. The geologists also found a second seamount nearby, which seems to have subsided faster, and which they have named "FitzRoy." The number of rounded pebbles and cobbles they dredged up suggest to these new surveyors of buried coasts that many other islands lie drowned between the Galápagos and the continent.

All these volcanoes, the new ones above the sea and the old ones below, were made when lava welled up through the crust of the sea floor. They mark a hot spot in the earth, where lava is always welling up. As the crust drifts eastward across this hot spot, it carries the old islands away, and new islands form in their place. The youngest islands in the Galápagos are in the east, the oldest in the west; Daphne lies precisely in the middle.

They rise, they are discovered by seeds and birds, they support Darwinian chains of action and reaction, and they sink again to the bottom of the sea, while new islands rise in their place. This rise and fall may have gone on here in the middle of the sea for as many as eighty or ninety million years.

The sight of Daphne Major conveys something like this to us, even in the first glance over the water, or in the last, as it revolves like a wood chip in the wake of a boat. We know we are looking at a place that was here before we came and will remain when we are gone. The very island will sink someday, and another will rise when it is drowned. The seasons will go on changing, and the cactus will suffer them. The waves will go on beating, and the cliffs will suffer them. Darwin's finches will keep their covenant with Darwin's islands, witnessed by a heap of stones.

Bibliography

Abbott, Ian. 1972. "The Ecology and Evolution of Passerine Birds on Islands." Dissertation, Monash University.

Abbott, Ian, L. K. Abbott, and Peter R. Grant. 1977. "Comparative Ecology of Galápagos Ground Finches (*Geospiza* Gould): Evaluation of the Importance of Floristic Diversity and Interspecific Competition." *Ecological Monographs* 47:151–84.

Anderson, E. 1948. "Hybridization of the Habitat." *Evolution* 2:1–9.

Anderson, E., and G. L. Stebbins, Jr. 1954. "Hybridization as an Evolutionary Stimulus." *Evolution* 8:378–88.

Atkins, Sir Hedley. 1974. *Down: The Home of the Darwins*. London: Curwen Press.

Averill, Anne L., and Ronald J. Prokopy. 1987. "Intraspecific Competition in the Tephritid Fruit Fly, *Rhagoletis pomonella*." *Ecology* 68:878–86.

Baker, Allan J. 1980. "Morphometric Differentiation in New Zealand Populations of the House Sparrow (*Passer domesticus*)." *Evolution* 34:638–53.

Bakun, Andrew. 1990. "Global Climate Change and Intensification of Coastal Ocean Upwelling." *Science* 247:198–201.

Bangham, Charles R. M., and Andrew J. McMichael. 1990. "Why the Long Latent Period?" *Nature* 348:388.

Barbosa, Pedro, and Jack C. Schultz, eds. 1987. *Insect Outbreaks*. San Diego: Academic Press, Inc.

Beebe, William. 1924. *Galápagos: World's End*. New York: G. P. Putnam's Sons.

Beer, Gillian. 1985. "Darwin's Reading and the Fictions of Development." In *The Darwinian Heritage. See Kohn*, ed. 1985, 543–88.

Benkman, Craig W., and Anna K. Lindholm. 1991. "The Advantages and Evolution of Morphological Novelty." *Nature* 349:519–20.

Berry, R. J. 1990. "Industrial Melanism and Peppered Moths (*Biston betularia* [L.])." *Biological Journal of the Linnean Society* 39:302–22.

Berthold, P., et al. 1992. "Rapid Microevolution of Migratory Behaviour in a Wild Bird Species." *Nature* 360:668–70.

Beverly, Stephen M., and Allan C. Wilson. 1985. "Ancient Origin for Hawaiian Drosophilinae Inferred from Protein Comparisons." *Proceedings of the National Academy of Sciences* 82:4753–57.

Bishop, J. A., and Laurence M. Cook. 1975. "Moths, Melanism and Clean Air." *Scientific American* 232 (January): 90–99.

Bloom, Barry R. 1992. "Back to a Frightening Future." *Nature* 358:538–64.

Bloom, Barry R., and Christopher J. L. Murray. 1992. "Tuberculosis: Commentary on a Reemergent Killer." *Science* 257:1055–64.

Boag, Peter T. 1983. "Galápagos Evolution Continues." *Nature* 301:12.

Boag, Peter T., and Peter R. Grant. 1978. "Heritability of External Morphology in Darwin's Finches." *Nature* 274:793–94.

———. 1981. "Intense Natural Selection in a Population of Darwin's Finches (*Geospizinae*) in the Galápagos." *Science* 214:82–85.

———. 1984. "The Classical Case of Character Release: Darwin's Finches (*Geospiza*) on Isla Daphne Major, Galápagos." *Biological Journal of the Linnean Society* 22:243–87.

———. 1984. "Darwin's Finches (*Geospiza*) on Isla Daphne Major, Galápagos: Breeding and Feeding Ecology in a Climatically Variable Environment." *Ecological Monographs* 54:463–89.

Bonner, John Tyler. 1980. *The Evolution of Culture in Animals.* Princeton: Princeton University Press.

———. 1993. *Life Cycles.* Princeton: Princeton University Press.

Bowman, Robert I. 1963. "Evolutionary Patterns in Darwin's Finches." *Occasional Papers of the California Academy of Sciences* 44:107–40.

———. 1983. "The Evolution of Song in Darwin's Finches." In *Patterns of Evolution. See* Bowman et al., eds. 1983, 237–325.

Bowman, Robert I., et al., eds. 1983. *Patterns of Evolution in Galápagos Organisms.* San Francisco: American Association for the Advancement of Science, Pacific Division.

Brakefield, Paul M. 1987. "Industrial Melanism: Do We Have the Answers?" *Trends in Ecology and Evolution* 2:117–22.

———. 1990. "A Decline of Melanism in the Peppered Moth, *Biston betularia*, in the Netherlands." *Biological Journal of the Linnean Society* 39:327–34.

Brookfield, J. F. Y. 1991. "The Resistance Movement." *Nature* 350:107–8.

Brown, Andrew J. Leigh. 1989. "Population Genetics at the DNA Level." *Oxford Surveys in Evolutionary Biology* 6:207–42.

Brown, W. L., and E. O. Wilson. 1956. "Character Displacement." *Systematic Zoology* 5:49–64.

Brussard, Peter F., ed. 1978. *Ecological Genetics: The Interface.* New York: Springer-Verlag.

Bumpus, Hermon Carey. 1899. "The Elimination of the Unfit as Illustrated by the Introduced Sparrow, *Passer domesticus.*" Woods Hole, Mass.: Biological Lectures, Marine Biological Laboratory: 209–26.

Bumpus, Herman Carey, Jr. 1947. *Hermon Carey Bumpus.* Minneapolis: University of Minnesota Press.

Burnet, F. M. 1962. "Evolution Made Visible: Current Changes in the Pattern of Disease." In *The Evolution of Living Organisms. See* Leeper, ed. 1962, 23–32.

Bush, Guy L., et al. 1989. "Sympatric Origins of *R. pomonella.*" *Nature* 339:346.

Cairns, John, Julie Overbaugh, and Stephan Miller. 1988. "The Origin of Mutants." *Nature* 335:142–45.

Carroll, Scott P., and Christin Boyd. 1992. "Host Race Radiation in the Soapberry Bug: Natural History with the History." *Evolution* 46:1052–69.

Carson, H. L. 1978. "Speciation and Sexual Selection in Hawaiian Drosophila." In *Ecological Genetics. See* Brussard, ed. 1978, 93–107.

———. 1992. "The Galápagos That Were." *Nature* 355:202–3.

Carson, H. L., Linda S. Chang, and Terrence W. Lyttle. 1982. "Decay of Female Sexual Behavior under Parthenogenesis." *Science* 218:68–70.

Caugant, Dominique, Bruce R. Levin, and Robert K. Selander. 1981. "Genetic Diversity and Temporal Variation in the *E. coli* Population of a Human Host." *Genetics* 98:467–90.

————. 1984. "Distribution of Multilocus Genotypes of *Escherichia coli* Within and Between Host Families." *Journal of Hygiene* 92:377–84.

Cayot, Linda J. 1985. "Effects of El Niño on Giant Tortoises and Their Environment." In *El Niño. See* Robinson and del Pino, eds. 1985, 363–98.

Chang, C. P., and F. W. Plapp, Jr. 1983. "DDT and Pyrethroids: Receptor Binding and Mode of Action in the House Fly." *Pesticide Biochemistry and Physiology* 20:76–85.

Charlesworth, Brian. 1990. "Life and Times of the Guppy." *Nature* 346:313–15.

Christie, D. M., et al. 1992. "Drowned Islands Downstream from the Galápagos Hotspot Imply Extended Speciation Times." *Nature* 355:246–48.

Clarke, Cyril A., Frieda M. M. Clarke, and H. C. Dawkins. 1990. "*Biston betularia* (the Peppered Moth) in West Kirby, Wirral, 1959–1989." *Biological Journal of the Linnean Society* 39:323–26.

Cohen, Mitchell L. 1992. "Epidemiology of Drug Resistance: Implications for a Post-Antimicrobial Era." *Science* 257:1050–55.

Conant, Sheila. 1988. "Geographic Variation in the Laysan Finch (*Telespyza cantans*)." *Evolutionary Ecology* 2:270–82.

————. 1988. "Saving Endangered Species by Translocation." *BioScience* 38:254–57.

Connell, Joseph H. 1980. "Diversity and the Coevolution of Competitors, or the Ghost of Competition Past." *Oikos* 35:131–38.

Conze, Edward, ed. 1971. *Buddhist Scriptures.* Harmondsworth, Middlesex: Penguin Books.

Cooke, F., and P. A. Buckley, eds. 1987. *Avian Genetics: A Population and Ecological Approach.* New York: Academic Press.

Creed, E. R. 1971. "Industrial Melanism in the Two-Spot Ladybird and Smoke Abatement." *Evolution* 25:290–93.

Creed, Robert, ed. 1971. *Ecological Genetics and Evolution.* Oxford, U.K.: Blackwell Scientific Publications.

Culliney, John L. 1988. *Islands in a Far Sea.* San Francisco: Sierra Club Books.

Curry, Robert L. 1985. "Breeding and Survival of Galápagos Mockingbirds during El Niño." In *El Niño. See* Robinson and del Pino, eds. 1985, 449–71.

Curry, Robert L., and Peter R. Grant. 1989. "Demography of the Cooperatively Breeding Galápagos Mockingbird, *Nesomimus parvulus*, in a Climatically Variable Environment." *Journal of Animal Ecology* 58:441–63.

Darwin, Charles R. 1987–89. *The Works of Charles Darwin.* 29 vols. Eds. Paul H. Barrett and R. B. Freeman. New York: New York University Press.

————. 1851–54. *A Monograph on the Sub-class Cirripedia.* London: Ray Society.

————. 1876. *Geological Observations on the Volcanic Islands and Parts of South America Visited During the Voyage of H.M.S. Beagle.* 2nd ed. London: Smith, Elder.

————. 1879. *A Naturalist's Voyage: Journal of Researches into the Natural History and Geology of the Countries Visited During the Voyage of H.M.S. Beagle Round the World, Under the Command of Capt. FitzRoy, R.N.* 2nd ed. London: John Murray.

————. 1958. *The Autobiography of Charles Darwin and Selected Letters.* Ed. Francis Darwin. New York: Dover Publications.

————. 1964 (1859). *On the Origin of Species.* Ed. Ernst Mayr. Facsimile of 1st ed. Cambridge, Mass.: Harvard University Press.

———. 1975. *Charles Darwin's Natural Selection*. Ed. R. C. Stauffer. Cambridge, U.K.: Cambridge University Press.

———. 1977. *The Collected Papers of Charles Darwin*. Ed. Paul H. Barrett. Chicago: University of Chicago Press.

———. 1981 (1871). *The Descent of Man, and Selection in Relation to Sex*. Princeton: Princeton University Press.

———. 1985–. *The Correspondence of Charles Darwin*. 8 vols. to date. Eds. Frederick Burkhardt and Sydney Smith. Cambridge, U.K.: Cambridge University Press.

———. 1987 (1836–44). *Charles Darwin's Notebooks*. Eds. Paul H. Barrett et al. Ithaca: Cornell University Press.

———. 1987. *Diary of the Voyage of the H.M.S. Beagle*. Vol. 1 in *Works. See* Darwin 1987–89.

———. 1987 (1839). *Journal of Researches*. Vols. 2 & 3 in *Works. See* Darwin 1987–89.

———. 1987. (1839–43). *The Zoology of the Voyage of H.M.S. Beagle*. Vols. 4–6 in *Works. See* Darwin 1987–89.

———. 1988 (1876). *On the Origin of Species*. 6th ed. Vol. 16 in *Works. See* Darwin 1987–89.

———. 1988 (1875). *The Variation of Animals and Plants under Domestication*. 2nd ed. Vols. 19–20 in *Works. See* Darwin 1987–89.

Dawkins, Richard. 1985. *The Blind Watchmaker*. New York: W. W. Norton.

———. 1991. "Darwin Triumphant." See Robinson and Tiger, eds., 23–39.

DeBenedictis, Paul A. 1968. "The Bill-Brace Feeding Behavior of the Galápagos Finch, *Geospiza conirostris*." *Condor* 68:206–8.

Desmond, Adrian. 1984. "Robert E. Grant: The Social Predicament of a Pre-Darwinian Transmutationist." *Journal of the History of Biology* 17:189–223.

Delbrück, Max. 1949. "A Physicist Looks at Biology." *Transactions of the Connecticut Academy of Arts and Sciences* 38:175–90.

Desmond, Adrian, and James Moore. 1991. *Darwin*. London: Michael Joseph.

Diamond, Jared M. 1987. "Learned Specializations of Birds." *Nature* 330:16–17.

Diamond, Jared M., and Ted J. Case, eds. 1986. *Community Ecology*. New York: Harper & Row.

Diehl, Scott Raymond. 1984. "The Role of Host Plant Shifts in the Ecology and Speciation of *Rhagoletis* Flies (Diptera: Tephritidae)." Dissertation, University of Texas at Austin.

Dillard, Annie. 1989. *The Writing Life*. New York: Harper & Row.

Dobson, Andrew P., et al. 1992. "Conservation Biology: The Ecology and Genetics of Endangered Species." In Berry, R. J., et al., eds. 1992. *Genes in Ecology*. Oxford, U.K.: Blackwell Scientific Publications, 405–30.

Dobzhansky, Theodosius, and Olga Pavlovsky. 1966. "Spontaneous Origin of an Incipient Species in the *Drosophila paulistorum* Complex." *Proceedings of the National Academy of Sciences* 55:727–33.

Dobzhansky, Theodosius, Olga Pavlovsky, and J. R. Powell. 1976. "Partially Successful Attempt to Enhance Reproductive Isolation between Semispecies of *Drosophila paulistorum*." *Evolution* 30:201–12.

Dobzhansky, Theodosius, and Boris Spassky. 1959. "*Drosophila paulistorum*, a Cluster of Species *in statu nascendi*." *Proceedings of the National Academy of Sciences* 45:419–28.

Dominey, Wallace J. 1984. "Effects of Sexual Selection and Life History on Speciation: Species Flocks in African Cichlids and Hawaiian Drosophila." In *Evolution of Fish Species Flocks. See* Echelle and Kornfield, eds. 1984, 231–49.

Dowdeswell, W. H. 1963. *The Mechanism of Evolution.* 3rd ed. London: Heinemann.

———. 1971. "Ecological Genetics and Biology Teaching." In *Ecological Genetics. See* Creed, ed. 1971, 363–78.

Echelle, Anthony A., and Irv Kornfield, eds. 1984. *Evolution of Fish Species Flocks.* Orono: University of Maine at Orono Press.

Ehrlich, Paul, and Anne Ehrlich. 1981. *Extinction.* New York: Random House.

Ehrlich, Paul R., Richard W. Holm, and Dennis R. Parnell. 1974. *The Processes of Evolution.* New York: McGraw-Hill.

Emerson, Ralph Waldo. 1983. "Spiritual Laws." In *Essays and Lectures.* Ed. Joel Porte, *The Library of America* 15. New York: Library of America. 305–23.

Endler, John A. 1977. *Geographic Variation, Speciation, and Clines.* Princeton: Princeton University Press.

———. 1978. "A Predator's View of Animal Color Patterns." *Evolutionary Biology* 11:319–64.

———. 1980. "Natural Selection on Color Patterns in *Poecilia reticulata.*" *Evolution* 34:76–91.

———. 1982. "Convergent and Divergent Effects of Natural Selection on Color Patterns in Two Fish Faunas." *Evolution* 36:178–88.

———. 1983. "Natural and Sexual Selection on Color Patterns in Poeciliid Fishes." *Environmental Biology of Fishes* 9:173–90.

———. 1986. *Natural Selection in the Wild.* Princeton: Princeton University Press.

———. 1986. "The Newer Synthesis? Some Conceptual Problems in Evolutionary Biology." *Oxford Surveys in Evolutionary Biology* 3:224–43.

———. 1988. "Sexual Selection and Predation Risk in Guppies." *Nature* 332:593–94.

———. 1989. "Conceptual and Other Problems in Speciation." In *Speciation. See* Otte and Endler, eds. 1989, 625–48.

Endler, John A., and Tracy McLellan. 1988. "The Processes of Evolution: Toward a Newer Synthesis." *Annual Review of Ecology and Systematics* 19:395–421.

Evans, L. T. 1984. "Darwin's Use of the Analogy between Artificial and Natural Selection." *Journal of the History of Biology* 17:113–40.

Feder, Jeffrey L. 1989. "The Biochemical Genetics of Host Race Formation and Sympatric Speciation in *Rhagoletis pomonella* (Diptera: Tephritidae)." Dissertation, Michigan State University.

———. 1990. "The Ecology and Genetics of Host Race Formation in *Rhagoletis pomonella.*" Research proposal, Princeton University.

Feder, Jeffrey L., and Guy L. Bush. 1989. "A Field Test of Differential Host-Plant Usage between Two Sibling Species of *Rhagoletis pomonella* Fruit Flies (Diptera: Tephritidae) and Its Consequences for Sympatric Models of Speciation." *Evolution* 43:1813–19.

———. 1989. "Gene Frequency Clines for Host Races of *Rhagoletis pomonella* in the Midwestern United States." *Heredity* 63:245–66.

Feder, Jeffrey L., Charles A. Chilcote, and Guy L. Bush. 1988. "Genetic Differentiation between Sympatric Host Races of the Apple Maggot Fly, *Rhagoletis pomonella.*" *Nature* 336:61–64.

———. 1989. "Are the Apple Maggot, *Rhagoletis pomonella,* and Blueberry Maggot, *R. mendax,* Distinct Species? Implications for Sympatric Speciation." *Entomologia Experimentalis et Applicata* 51:113–23.

———. 1990. "The Geographic Pattern of Genetic Differentiation between Host Associated Populations of *Rhagoletis pomonella* (Diptera: Tephritidae) in the Eastern United States and Canada." *Evolution* 44:570–94.

————. 1990. "Regional, Local and Microgeographic Allele Frequency Variation between Apple and Hawthorn Populations of *Rhagoletis pomonella* in Western Michigan." *Evolution* 44:595–608.

Fiorito, Graziano, and Pietro Scotto. 1992. "Observational Learning in *Octopus vulgaris*." *Science* 256:545–47.

FitzRoy, Robert. 1839. *Narrative of the Surveying Voyages of His Majesty's Ships Adventure and Beagle, between the Years 1826 and 1836, Describing Their Examination of the Southern Shores of South America, and the Beagle's Circumnavigation of the Globe.* 3 vols. and app. London: Henry Colburn.

Fleischer, Robert C., Sheila Conant, and Marie P. Morin. 1991. "Genetic Variation in Native and Translocated Populations of the Laysan Finch (*Telespiza cantans*)." *Heredity* 66:125–30.

Fleischer, Robert C., and Richard F. Johnston. 1982. "Natural Selection on Body Size and Proportions in House Sparrows." *Nature* 298:747–49.

————. 1984. "The Relationship between Winter Climate and Selection on Body Size of House Sparrows." *Canadian Journal of Zoology* 62:405–10.

Ford, E. B. 1960. "Evolution in Progress." In *Evolution after Darwin. See* Tax, ed. 1980, 181–96.

————. 1975. *Ecological Genetics.* 4th ed. London: Chapman & Hall.

Freed, Leonard A., Sheila Conant, and Robert C. Fleischer. 1987. "Evolutionary Ecology and Radiation of Hawaiian Passerine Birds." *Trends in Ecology and Evolution* 2:196–203.

Futuyma, Douglas J. 1986. *Evolutionary Biology.* 2nd ed. Sunderland, Mass.: Sinauer Associates.

Gao, Feng, et al. 1992. "Human Infection by Genetically Diverse SIV SM-Related HIV-2 in West Africa." *Nature* 358:495–99.

Garcia-Bustos, Jose, and Alexander Tomasz. 1990. "A Biological Price of Antibiotic Resistance: Major Changes in the Peptidoglycan Structure of Penicillin-Resistant Pneumococci." *Proceedings of the National Academy of Sciences* 87:5415–19.

Gelter, Hans P., H. Lisle Gibbs, and Peter T. Boag. "Large Deletions in the Control Region of Darwin's Finch Mitochondrial DNA: Evolutionary and Functional Implications." *Proceedings of the National Academy of Sciences*, in press.

Georghiou, George P. 1986. "The Magnitude of the Resistance Problem." In *Pesticide Resistance. See* Roush and Tabashnik, eds., 1986, 14–43.

Gibbons, Ann. 1992. "Exploring New Strategies to Fight Drug-Resistant Microbes." *Science* 257:1036–38.

Gibbs, H. Lisle. 1988. "Heritability and Selection on Clutch Size in Darwin's Medium Ground Finches (*Geospiza fortis*)." *Evolution* 42:750–62.

Gibbs, H. Lisle, and Peter R. Grant. 1987. "Adult Survivorship in Darwin's Ground Finch (*Geospiza*) Populations in a Variable Environment." *Journal of Animal Ecology* 56:797–813.

————. 1987. "Ecological Consequences of an Exceptionally Strong El Niño Event on Darwin's Finches." *Ecology* 68:1735–46.

————. 1987. "Oscillating Selection on Darwin's Finches." *Nature* 327:511–13.

————. 1989. "Inbreeding in Darwin's Medium Ground Finches (*Geospiza fortis*)." *Evolution* 43:1273–84.

Gibbs, H. Lisle, Peter R. Grant, and Jon Weiland. 1984. "Breeding of Darwin's Finches at an Unusually Early Age in an El Niño Year." *Auk* 101:873–74.

Gill, Frank B. 1980. "Historical Aspects of Hybridization between Blue-Winged and Golden-Winged Warblers." *Auk* 97:1–18.

————. 1989. *Ornithology.* New York: W. H. Freeman.

Gillespie, Neal C. 1979. *Charles Darwin and the Problem of Creation.* Chicago: University of Chicago Press.

Gingerich, Philip D. 1983. "Rates of Evolution: Effects of Time and Temporal Scaling." *Science* 222:159–61.

Gish, Duane T. 1979. *Evolution? The Fossils Say No!* San Diego: Creation-Life Publishers.

Godard, R. 1991. "Long-Term Memory of Individual Neighbours in a Migratory Songbird." *Nature* 350:228–29.

Gorman, Owen T., et al. 1990. "Evolution of the Nucleoprotein Gene of Influenza A Virus." *Journal of Virology* 64:1487–97.

Gould, Fred. 1991. "The Evolutionary Potential of Crop Pests." *American Scientist* 79:496–507.

Gould, James L., and Carol Grant Gould. 1989. *Sexual Selection.* New York: Scientific American Library.

Gould, Stephen Jay. 1983. *Hen's Teeth and Horses' Toes.* New York: W. W. Norton.

————. 1989. *Wonderful Life.* New York: W. W. Norton.

Grant, B. Rosemary. 1985. "Selection on Bill Characters in a Population of Darwin's Finches: *Geospiza conirostris* on Isla Genovesa, Galápagos." *Evolution* 39:523–32.

Grant, B. Rosemary, and Peter R. Grant. 1979. "The Feeding Ecology of Darwin's Ground Finches." *Noticias de Galápagos* 14–18.

————. 1981. "Exploitation of *Opuntia* Cactus by Birds on the Galápagos." *Oecologia* 49:179–87.

————. 1982. "Niche Shifts and Competition in Darwin's Finches: *Geospiza conirostris* and Congeners." *Evolution* 36:637–57.

————. 1983. "Fission and Fusion in a Population of Darwin's Finches: An Example of the Value of Studying Individuals in Ecology." *Oikos* 41:530–47.

————. 1989. *Evolutionary Dynamics of a Natural Population.* Princeton: Princeton University Press.

————. 1989. "Natural Selection in a Population of Darwin's Finches." *American Naturalist* 133:377–93.

————. 1993. "Evolution of Darwin's Finches Caused by a Rare Climatic Event." *Proceedings of the Royal Society of London (B)* 251:111–17.

Grant, Bruce, and Rory J. Howlett. 1988. "Background Selection by the Peppered Moth (*Biston betularia* Linn.): Individual Differences." *Biological Journal of the Linnean Society* 33:217–32.

Grant, Peter R. 1966. "Ecological Compatibility of Bird Species on Islands." *American Naturalist* 100:451–62.

————. 1966. "Late Breeding on the Tres Marías Islands." *Condor* 68:249–52.

————. 1970. "Variation and Niche Width Reexamined." *American Naturalist* 104:589–90.

————. 1972. "Centripetal Selection and the House Sparrow." *Systematic Zoology* 21:23–30.

————. 1972. "Convergent and Divergent Character Displacement." *Biological Journal of the Linnean Society* 4:39–68.

————. 1972. "Interspecific Competition among Rodents." *Annual Review of Ecology and Systematics* 3:79–106.

————. 1975. "The Classical Case of Character Displacement." *Evolutionary Biology* 8:237–337.

————. 1977. "Review of D. Lack, 1976, *Island Biology.*" *Bird-Banding* 48:296–300.

————. 1981. "The Feeding of Darwin's Finches on *Tribulus cistoides* (L.) Seeds." *Animal Behaviour* 29:785–93.

————. 1981. "Speciation and the Adaptive Radiation of Darwin's Finches." *American Scientist* 69:653–63.

————. 1986. *Ecology and Evolution of Darwin's Finches.* Princeton: Princeton University Press. (The best-thumbed book in this bibliography.)

————. 1986. "Interspecific Competition in Fluctuating Environments." In *Community Ecology. See* Diamond and Case, eds. 1986, 173–91.

————. 1993. "Hybridization of Darwin's Finches on Isla Daphne Major, Galápagos." *Philosophical Transactions of the Royal Society of London (B)* 340:127–39.

Grant, Peter R., and Peter T. Boag. 1980. "Rainfall on the Galápagos and the Demography of Darwin's Finches." *Auk* 97:227–44.

Grant, Peter R., and B. Rosemary Grant. 1985. "Responses of Darwin's Finches to Unusual Rainfall." In *El Niño. See* Robinson and del Pino, eds. 1985, 417–47.

————. 1989. "The Slow Recovery of *Opuntia megasperma* on Española." *Noticias de Galápagos* 48:13–15.

————. 1989. "Sympatric Speciation and Darwin's Finches." In *Speciation. See* Otte and Endler, eds. 1989, 433–57.

————. 1992. "Demography and the Genetically Effective Sizes of Two Populations of Darwin's Finches." *Ecology* 73:766–84.

————. 1992. "Global Warming and the Galápagos." *Noticias de Galápagos* 51:14–16.

————. 1992. "Hybridization of Bird Species." *Science* 256:193–97.

Grant, Peter, K. Thalia Grant, and B. Rosemary Grant. 1991. "*Erythrina velutina* and the Colonization of Remote Islands." *Noticias de Galápagos* 3–5.

Grant, P. R., and Nicola Grant. 1979. "Breeding and Feeding of Galápagos Mockingbirds, *Nesomimus parvulus.*" *Auk* 96:723–35.

Grant, Peter R., and Henry S. Horn, eds. 1992. *Molds, Molecules, and Metazoa.* Princeton: Princeton University Press.

Grant, P. R., and T. D. Price. 1981. "Population Variation in Continuously Varying Traits as an Ecological Genetics Problem." *American Zoologist* 21:795–811.

Grant, P. R., et al. 1975. "Finch Numbers, Owl Predation and Plant Dispersal on Isla Daphne Major, Galápagos." *Biological Journal of the Linnean Society* 19:239–57.

————. 1976. "Darwin's Finches: Population Variation and Natural Selection." *Proceedings of the National Academy of Sciences* 73:257–61.

————. 1985. "Variation in the Size and Shape of Darwin's Finches." *Biological Journal of the Linnean Society* 25:1–39.

Greene, John C. 1959. *The Death of Adam.* Ames: Iowa State University Press.

Greenwood, Jeremy J. D. 1990. "Changing Migration Behaviour." *Nature* 345:209–10.

————. 1993. "Theory Fits the Bill in the Galápagos Islands." *Nature* 362:699.

Gruson, Lindsey. 1992. "Throwing Back Undersize Fish Is Said to Encourage Smaller Fry." *New York Times*, January 7: C4.

Gustafsson, Lars, and Tomas Pärt. 1990. "Acceleration of Senescence in the Collared Flycatcher *Ficedula albicollis* by Reproductive Costs." *Nature* 347:279–81.

Gustafsson, Lars, and William J. Sutherland. 1988. "The Costs of Reproduction in the Collared Flycatcher *Ficedula albicollis.*" *Nature* 335:813–15.

Hahn, Beatrice H., et al. 1986. "Genetic Variation in HTLV–III/LAV over Time in Patients with AIDS or at Risk for AIDS." *Science* 232:1548–53.

Haldane, J. B. S. 1949. "Human Evolution: Past and Future." In *Genetics. See* Jepson, Simpson, and Mayr, eds. 1949, 405–18.

———. 1949. "Suggestions as to the Quantitative Measurement of Rates of Evolution." *Evolution* 3:51–56.

Hall, G. A., et al. 1986. "Effects of El Niño–Southern Oscillation (ENSO) on Terrestrial Birds." *International Ornithological Congress* (19th): 1759–69.

Hall, Linda M., and Durgadas P. Kasbekar. 1989. "Drosophila Sodium Channel Mutations Affect Pyrethroid Sensitivity." In Narahashi, Toshio, and Janice E. Chambers, eds. 1989. *Insecticide Action*. New York: Plenum, 99–114.

Harris, J. Arthur. 1911. "A Neglected Paper on Natural Selection in the English Sparrow." *American Naturalist* 45:314–18.

Harris, Lester E., Jr. 1976. *Galápagos*. Nashville: Southern Publishing Association.

Harris, Michael P. 1974. *A Field Guide to the Birds of Galápagos*. London: Collins.

Harrison, R. G. 1978. "Ecological Parameters and Speciation in Field Crickets." In *Ecological Genetics. See* Brussard, ed. 1978, 145–58.

Hillis, David M., et al. 1992. "Experimental Phylogenetics: Generation of a Known Phylogeny." *Science* 255:589–92.

Hochberg, Michael E., and John H. Lawton. 1990. "Competition Between Kingdoms." *Trends in Ecology and Evolution* 5:367–71.

Holt, Robert. 1990. "Birds under Selection. Review of *Evolutionary Dynamics of a Natural Population*, by B. Rosemary Grant and Peter R. Grant." *Science* 249:306–7.

———. 1990. "The Microevolutionary Consequences of Climate Change." *Trends in Ecology and Evolution* 5:311–15.

Houde, Anne E., and John A. Endler. 1990. "Correlated Evolution of Female Mating Preferences and Male Color Patterns in the Guppy, *Poecilia reticulata*." *Nature* 248:1405–8.

"A Howling Blizzard." 1898. Providence *Journal*, February 1:1.

Hughes, Walter T. 1988. "A Tribute to Toilet Paper." *Reviews of Infectious Diseases* 10:218–22.

Huxley, Thomas Henry. 1893. *Darwiniana*. New York: D. Appleton.

———. 1968 (1863). *On the Origin of Species, or, The Causes of the Phenomena of Organic Nature*. Ann Arbor: University of Michigan Press.

———. 1989 (1894). *Evolution and Ethics*. Eds. James Paradis and George C. Williams. Princeton: Princeton University Press.

Jackson, Michael H. 1985. *Galápagos*. Calgary, Alberta.: University of Calgary Press.

Jepson, Glenn L., George Gaylord Simpson, and Ernst Mayr, eds. 1949. *Genetics, Paleontology, and Evolution*. Princeton: Princeton University Press.

Johnson, Phillip E. 1991. *Darwin on Trial*. Washington, D.C.: Regnery Gateway.

Jones, J. S. 1981. "Models of Speciation—The Evidence from Drosophila." *Nature* 289:743–44.

———. 1982. "St. Patrick and the Bacteria." *Nature* 296:113–14.

Kaneshiro, Kenneth Y. 1988. "Speciation in the Hawaiian Drosophila." *BioScience* 38:258–63.

Keeton, William T., and James L. Gould. 1986. *Biological Science*. 4th ed. New York: W. W. Norton.

Kendrick, Amrit Work. 1988. "Santa Cruz Fact Sheet." *Noticias de Galápagos* 46:5–7.

Kettlewell, H. B. D. 1958. "A Survey of the Frequencies of *Biston betularia* (L.) (Lep.) and Its Melanic Forms in Great Britain." *Heredity* 12:51–72.

Kettlewell, Bernard. 1973. *The Evolution of Melanism*. Oxford, U.K.: Clarendon Press.

Kingsland, Sharon. 1970. "David Lambert Lack." *Dictionary of Scientific Biography*, 521–23.

———. 1985. *Modeling Nature: Episodes in the History of Population Ecology*. Chicago: University of Chicago Press.

Kofahl, Robert E. 1977. *Handy-Dandy Evolution Refuter*. San Diego: Beta Books.

Kohn, David, ed. 1985. *The Darwinian Heritage: A Centennial Retrospect*. Princeton: Princeton University Press.

———. 1985. "Darwin's Principle of Divergence as Internal Dialogue." In *Darwinian Heritage*. See Kohn, ed. 1985, 245–57.

Koshland, Daniel E., Jr. 1992. "The Microbial Wars." *Science* 257:1021.

Köster, Friedemann, and Heide Köster. 1983. "Twelve Days among the 'Vampire Finches' of Wolf Island." *Noticias de Galápagos* 38:4–10.

Kramer, P. 1984. "Man and Other Introduced Organisms." *Biological Journal of the Linnean Society* 21:253–58.

Krause, Richard M. 1992. "The Origin of Plagues: Old and New." *Science* 257:1073–78.

Krebs, John R. 1991. "The Case of the Curious Bill." *Nature* 349:465.

Krieber, Michel, and Michael R. Rose. 1986. "Molecular Aspects of the Species Barrier." *Annual Review of Ecology and Systematics* 17:465–85.

Lacey, R. W. 1984. "Evolution of Microorganisms and Antibiotic Resistance." *The Lancet*: 1022–25.

Lack, David. 1940. "Evolution of the Galápagos Finches." *Nature* 146:324–27.

———. 1945. The Galápagos Finches (*Geospizinae*): A Study in Variation. *Occasional Papers of the California Academy of Sciences* 21:1–159.

———. 1964. "Darwin's Finches." In *A New Dictionary of Birds*. Ed. Sir A. Landsborough Thomson. London: Thomas Nelson & Sons. 178–79.

———. 1968. *Ecological Adaptations for Breeding in Birds*. London: Methuen.

———. 1973. "My Life as an Amateur Ornithologist." *Ibis* 115:421–31.

———. 1983 (1947). *Darwin's Finches*. Eds. Laurene M. Ratcliffe and Peter T. Boag. Cambridge, U.K.: Cambridge University Press.

Laurie, Andrew. 1983. "Marine Iguanas Suffer as El Niño Breaks All Records." *Noticias de Galápagos* 38:11.

Leakey, Richard E. 1981. *The Making of Mankind*. New York: E. P. Dutton.

Leeper, G. W., ed. 1962. *The Evolution of Living Organisms*. Melbourne: Melbourne University Press.

Levin, Bruce R., Dominique A. Caugant, and Robert K. Selander. 1991. "The Genetic Response of the Human *E. coli* Flora to Antibiotic Treatment." Personal communication.

Levin, Donald A., ed. 1979. *Hybridization: An Evolutionary Perspective*. Stroudsburg, Pa.: Dowden, Hutchinson & Ross.

Levy, Avraham A., and Virginia Walbot. 1990. "Regulation of the Timing of Transposable Element Excision During Maize Development." *Science* 248:1534–37.

Levy, Stuart B. 1978. "Emergence of Antibiotic-Resistant Bacteria in the Intestinal Flora of Farm Inhabitants." *Journal of Infectious Diseases* 137:688–90.

Lewin, Roger. 1983. "Finches Show Competition in Ecology." *Science* 219:1411–12.

———. 1983. "Santa Rosalia Was a Goat." *Science* 221:636–39.

Lewontin, Richard C. 1974. *The Genetic Basis of Evolutionary Change*. New York: Columbia University Press.

————. 1978. "Adaptation." *Scientific American* 239:213–30.

————. 1982. *Human Diversity*. New York: Scientific American Books.

Lewontin, Richard C., and L. C. Birch. 1966. "Hybridization as a Source of Variation for Adaptation to New Environments." *Evolution* 20:315–36.

Loughney, Kate, Robert Kreber, and Barry Ganetzky. 1989. "Molecular Analysis of the *Para* Locus, a Sodium Channel Gene in Drosophila." *Cell* 58:1143–54.

Lowe, Percy R. 1936. "The Finches of the Galápagos in Relation to Darwin's Conception of Species." *Ibis* 13:310–21.

Lyell, Charles. 1990 (1830–1833). *The Principles of Geology*. Facsimile of the 1st ed. 3 vols. Chicago: University of Chicago Press.

Mallet, J. L. B. 1990. "Evolution of Insecticide Resistance." *Trends in Ecology and Evolution* 5:164–65.

Mani, G. S. 1990. "Theoretical Models of Melanism in *Biston betularia*—A Review." *Biological Journal of the Linnean Society* 39:355–71.

May, Robert M., and Andrew P. Dobson. 1986. "Population Dynamics and the Rate of Evolution of Pesticide Resistance." In *Pesticide Resistance. See* Roush and Tabashnik, eds. 1986, 170–93.

Mayr, Ernst. 1965. *Animal Species and Evolution*. Cambridge, Mass.: Belknap Press of Harvard University Press.

————. 1970. *Populations, Species and Evolution*. Cambridge, Mass.: Belknap Press of Harvard University Press.

————. 1982. *The Growth of Biological Thought*. Cambridge, Mass.: Belknap Press of Harvard University Press.

————. 1986. "The Contributions of Birds to Evolutionary Theory." *International Ornithological Congress* (19th): 2718–23.

————. 1991. *One Long Argument*. Cambridge, Mass.: Harvard University Press.

McDonald, John F. 1983. "The Molecular Basis of Adaptation: A Critical Review of Relevant Ideas and Observations." *Annual Review of Ecology and Systematics* 14:77–102.

Melville, Herman. 1987. *The Essential Melville*. New York: Ecco Press.

Merlen, Godfrey. 1985. "The Nature of El Niño: A Perspective." In *El Niño. See* Robinson and del Pino, eds. 1985, 133–50.

Miller, Julie Ann. 1989. "Diseases for Our Future." *BioScience* 39:509–17.

Millington, S. J., and Peter R. Grant. 1983. "Feeding Ecology and Territoriality of the Cactus Finch *Geospiza scandens* on Isla Daphne Major, Galápagos." *Oecologia* 58:76–83.

Millington, S. J., and Trevor D. Price. 1982. "Birds on Daphne Major 1979–1981." *Noticias de Galápagos* 35:25–27.

Milner, Richard. 1990. *The Encyclopedia of Evolution*. New York: Facts on File.

Milton, John. 1969 (1667–1674). *Paradise Lost, Paradise Regained and Samson Agonistes*. Garden City, N.Y.: Doubleday.

Moore, James R. 1985. "Darwin of Down: The Evolutionist as Squarson-Naturalist." In *Darwinian Heritage. See* Kohn, ed. 1985, 435–81.

Moorehead, Alan. 1971. *Darwin and the Beagle*. Reprint ed. Harmondsworth, Middlesex: Penguin Books.

Neu, Harold C. 1992. "The Crisis in Antibiotic Resistance." *Science* 257:1064–73.

Newton, Ian. 1973. *Finches*. New York: Taplinger Publishing.

Otte, Daniel. 1989. "Speciation in Hawaiian Crickets." In *Speciation. See* Otte and Endler, eds. 1989, 482–525.

Otte, Daniel, and John A. Endler, eds. 1989. *Speciation and Its Consequences.* Sunderland, Mass.: Sinauer Associates.

Parkin, David T. 1987. "Evolutionary Genetics of House Sparrows." In *Avian Genetics. See* Cooke and Buckley, eds. 1987, 381–406.

Patterson, Colin. 1978. *Evolution.* Ithaca: Cornell University Press.

Pearl, Raymond. 1911. "Data on the Relative Conspicuousness of Fowls." *American Naturalist* 45:107–17.

———. 1917. "The Selection Problem." *American Naturalist* 51:65–91.

———. 1930. "Requirements of a Proof That Natural Selection Has Altered a Race." *Scientia* 47:175–86.

Perrins, C. M., and T. R. Birkhead. 1983. *Avian Ecology.* Bishopbriggs, Glasgow: Blackie & Sons.

Pfeiffer, John E. 1982. *The Creative Explosion.* Ithaca: Cornell University Press.

———. 1985. *The Emergence of Humankind.* 4th ed. New York: Harper & Row.

Phillips, Rodney E., et al. 1991. "Human Immunodeficiency Virus: Genetic Variation That Can Escape Cytotoxic T Cell Recognition." *Nature* 354:453.

Plapp, Frederick W., Jr. 1986. "Genetics and Biochemistry of Insecticide Resistance in Arthropods." In *Pesticide Resistance. See* Roush and Tabashnik, eds. 1986, 74–85.

Plapp, Frederick W., Jr., et al. 1990. "Monitoring and Management of Pyrethroid Resistance in the Tobacco Budworm (Lepidoptera: Noctuidae) in Texas, Mississippi, Louisiana, Arkansas, and Oklahoma." *Journal of Economic Entomology* 83:335–41.

Plapp, Frederick W., Jr., et al. 1986. "Management of Pyrethroid-Resistant Tobacco Budworms on Cotton in the United States." In *Pesticide Resistance.* Roush and Tabashnik, eds. 1986, 237–60.

Porter, Duncan M. 1983. "Vascular Plants of the Galápagos: Origins and Dispersal." In *Patterns of Evolution. See* Bowman et al., eds. 1983, 33–96.

———. 1985. "The *Beagle* Collector and His Collections." In *The Darwinian Heritage. See* Kohn, ed. 1985, 973–1019.

———. 1987. "Darwin Notes on *Beagle* Plants." *Bulletin of the British Museum of Natural History* (historical ser.) 14:145–233.

Prevosti, Antonio, et al. 1985. "The Colonization of *Drosophila subobscura* in Chile. II. Clines in the Chromosomal Arrangements." *Evolution* 39:838–44.

———. 1990. "Clines of Chromosomal Arrangements of *Drosophila subobscura* in South America Evolve Closer to Old World Patterns." *Evolution* 44:218–21.

Price, Trevor D. 1984. "The Evolution of Sexual Size Dimorphism in Darwin's Finches." *American Naturalist* 123:500–18.

———. 1984. "Sexual Selection on Body Size, Territory, and Plumage Variables in a Population of Darwin's Finches." *Evolution* 38:327–41.

———. 1990. "Memoir of Life on Daphne Major." Unpublished.

Price, Trevor D., and Peter T. Boag. 1987. "Selection in Natural Populations of Birds." In *Avian Genetics. See* Cooke and Buckley, eds. 1987, 257–87.

Price, Trevor D., and Peter R. Grant. 1984. "Life History Traits and Natural Selection for Small Body Size in a Population of Darwin's Finches." *Evolution* 38:483–94.

Price, Trevor D., Peter R. Grant, and Peter T. Boag. 1984. "Genetic Changes in the Morphological Differentiation of Darwin's Ground Finches." In Wöhrmann, K., and V. Loeschcke, eds. 1984. *Population Biology and Evolution.* New York: Springer-Verlag, 49–66.

Price, Trevor D., et al. 1984. "Recurrent Patterns of Natural Selection in a Population of Darwin's Finches." *Nature* 309:787–89.

Prokopy, Ronald J., and Bernard D. Roitberg. 1984. "Foraging Behavior of True Fruit Flies." *American Scientist* 72:41–49.

Prokopy, Ronald J., et al. 1982. "Associative Learning in Egglaying Site Selection by Apple Maggot Flies." *Science* 218:76–77.

Provine, William B. 1985. "Adaptation and Mechanisms of Evolution after Darwin: A Study in Persistent Controversies." In *The Darwinian Heritage. See* Kohn, ed. 1985, 825–66.

———. 1993. "Scientific Supernaturalism. A Review of *The Origin of Species Revisited: The Theories of Evolution and of Abrupt Appearance,* by W. R. Bird." *Biology and Philosophy* 8:111–24.

Raimondi, Peter T. 1992. "Adult Plasticity and Rapid Larval Evolution in a Recently Isolated Barnacle Population." *Biological Bulletin* 182:210–20.

Ratcliffe, Laurene M., and Peter R. Grant. 1983. "Species Recognition in Darwin's Finches (*Geospiza* Gould). I. Discrimination by Morphological Cues." *Animal Behavior* 31:1139–53.

———. 1983. "Species Recognition in Darwin's Finches (*Geospiza* Gould). II. Geographic Variation in Mate Preference." *Animal Behaviour* 31:1154–65.

———. 1985. "Species Recognition in Darwin's Finches (*Geospiza* Gould). III. Male Responses to Playback of Different Song Types, Dialects and Heterospecific Songs." *Animal Behaviour* 33:290–307.

Ratner, Lee, et al. 1985. "Complete Nucleotide Sequence of the AIDS Virus, HTLV-III." *Nature* 313:277–84.

Raup, David M. 1991. *Extinction.* New York: W. W. Norton.

Raymond, Michel, et al. 1991. "Worldwide Migration of Amplified Insecticide Resistance Genes in Mosquitoes." *Nature* 350:151–53.

Reznick, David, and John A. Endler. 1982. "The Impact of Predation on Life History Evolution in Trinidadian Guppies (*Poecilia reticulata*)." *Evolution* 36:160–77.

Reznick, David A., Heather Bryga, and John A. Endler. 1990. "Experimentally Induced Life-History Evolution in a Natural Population." *Nature* 346:357–59.

Rheinberger, Hans-Jörg, and Peter McLaughlin. 1984. "Darwin's Experimental Natural History." *Journal of the History of Biology* 17:345–68.

Ricklefs, Robert E. *Ecology.* 1990. 3rd ed. New York: W. H. Freeman.

Ridley, Mark. 1985. *The Problems of Evolution.* Oxford, U.K.: Oxford University Press.

Riedl, Helmut. 1983. "Analysis of Codling Moth Phenology in Relation to Latitude, Climate and Food Availability." In Brown, V. K., and I. Hodek, eds. 1983. *Diapause and Life Cycle Strategies in Insects.* The Hague: Dr. W. Junk Publishers, 233–52.

Robinson, Gary, and Eugenia M. del Pino, eds. 1985. *El Niño in the Galápagos Islands: The 1982–1983 Event.* Quito, Ecuador: Fundación Charles Darwin para las Islas Galápagos.

Robinson, Michael H., and Lionel Tiger, eds. 1991. *Man and Beast Revisited.* Washington, D.C.: Smithsonian Institution Press.

Robson, G. C., and O. W. Richards. 1936. *The Variation of Animals in Nature.* London: Longmans, Green.

Roush, R. T., and B. Tabashnik, eds. 1986. *Pesticide Resistance in Arthropods.* New York: Chapman.

Ruse, Michael, ed. 1988. *But Is It Science?* Buffalo: Prometheus Books.

Salkoff, Lawrence, et al. 1987. "Molecular Biology of the Voltage-Gated Sodium Channel." *Trends in Neurosciences* 10:522–26.

Salvin, O. 1876. "On the Avifauna of the Galápagos Archipelago." *Transactions of the Zoological Society of London* 9:447–510.

Schluter, Dolph. 1982. "Distributions of Galápagos Ground Finches along an Altitudinal Gradient: The Importance of Food Supply." *American Naturalist* 63:1504–17.

———. 1982. "Seed and Patch Selection by Galápagos Ground Finches: Relation to Foraging Efficiency and Food Supply." *Ecology* 63:1106–20.

———. 1986. "Character Displacement Between Distantly Related Taxa? Finches and Bees in the Galápagos." *American Naturalist* 127:95–102.

———. 1986. "Morphological Adaptation and Diet in the Galápagos Ground Finches." *International Ornithological Congress* (19th), 2283–95.

———. 1988. "Character Displacement and the Adaptive Divergence of Finches on Islands and Continents." *American Naturalist* 131:799–824.

———. 1988. "Estimating the Form of Natural Selection on a Quantitative Trait." *Evolution* 42:849–61.

———. 1988. "The Evolution of Finch Communities on Islands and Continents: Kenya vs. Galápagos." *Ecological Monographs* 58:229–49.

Schluter, Dolph, and Peter R. Grant. 1982. "The Distribution of *Geospiza difficilis* in Relation to *G. fuliginosa* in the Galápagos Islands: Tests of Three Hypotheses." *Evolution* 36:1213–26.

———. 1984. "Determinants of Morphological Patterns in Communities of Darwin's Finches." *American Naturalist* 123:175–96.

———. 1984. "Ecological Correlates of Morphological Evolution in a Darwin's Finch, *Geospiza difficilis*." *Evolution* 38:856–69.

Schluter, Dolph, and J. Donald McPhail. 1992. "Ecological Character Displacement and Speciation in Sticklebacks." *American Naturalist* 140:85–108.

Schluter, Dolph, Trevor D. Price, and Peter R. Grant. 1985. "Ecological Character Displacement in Darwin's Finches." *Science* 227:1056–59.

Schluter, Dolph, Trevor D. Price, and Locke Rowe. 1991. "Conflicting Selection Pressures and Life History Trade-Offs." *Proceedings of the Royal Society of London (B)* 246:11–17.

Schluter, Dolph, and James N. M. Smith. 1986. "Natural Selection on Beak and Body Size in the Song Sparrow." *Evolution* 40:221–31.

Searle, Jeremy B. 1992. "When Is a Species Not a Species?" *Current Biology* 2:407–8.

Sheppard, Carol M. 1993. "Benjamin Walsh: First State Entomologist of Illinois and Proponent of Darwinian Theory." Unpublished.

Sheppard, P. M. 1967. *Natural Selection and Heredity.* 3rd ed. London: Hutchinson University Library.

Sibley, Charles G., and Jon E. Ahlquist. 1990. *Phylogeny and Classification of Birds.* New Haven: Yale University Press.

Simberloff, Daniel. 1984. "The Great God of Competition." *The Sciences* 24.4: 16–22.

Smith, G. T. Corley. 1987. "Looking Back." *Noticias de Galápagos* 45:11–16.

———. 1990. "A Brief History of the Charles Darwin Foundation for the Galápagos Islands 1959–1988." *Noticias de Galápagos* 49:4–36.

Smith, James N. M., and Hugh P. A. Sweatman. 1976. "Feeding Habits and Morphological Variation in Cocos Finches." *Condor* 78:244–48.

Smith, James N. M., et al. 1978. "Seasonal Variation in Feeding Habits of Darwin's Ground Finches." *Ecology* 59:1137–50.

Smith, R. C., et al. 1992. "Ozone Depletion: Ultraviolet Radiation and Phytoplankton Biology in Antarctic Waters." *Science* 255:952–59.

Sober, Elliott. 1984. *The Nature of Selection*. Cambridge, Mass.: MIT Press.

———. 1985. "Darwin on Natural Selection: A Philosophical Perspective." In *The Darwinian Heritage. See* Kohn, ed. 1985, 867–99.

Steadman, David W. 1982. "The Origin of Darwin's Finches (Fringillidae: Passeriformes)." *Transactions of the San Diego Society of Natural History* 19:279–96.

———. 1984. "The Status of *Geospiza magnirostris* on Isla Floreana, Galápagos." *Bulletin of the British Ornithological Club* 104:99–102.

———. 1986. "Holocene Terrestrial Gastropod Faunas from Isla Santa Cruz and Isla Floreana, Galápagos: Evidence for Late Holocene Declines." *Transactions of the San Diego Society of Natural History* 21:89–110.

———. 1986. *Holocene Vertebrate Fossils from Isla Floreana, Galápagos*. Smithsonian Contributions to Zoology. No. 413. Washington, D.C.: Smithsonian Institution Press. (My chief source for the story of *Magnirostris magnirostris*.)

Steadman, David W., and Steven Zousmer. 1988. *Galápagos*. Washington, D.C.: Smithsonian Institution Press.

Stone, Irving. 1980. *The Origin*. Garden City, N.Y.: Doubleday.

Stoppard, Tom. 1981. "This Other Eden." *Noticias de Galápagos* 34:6–7.

Strong, Donald R., Jr., Lee Ann Szyska, and Daniel S. Simberloff. 1979. "Tests of Community-Wide Character Displacement against Null Hypothesis." *Evolution* 33:897–913.

Strong, Donald, Jr., et al., eds. 1984. *Ecological Communities*. Princeton: Princeton University Press.

Sulloway, Frank J. 1982. "The *Beagle* Collections of Darwin's Finches (*Geospizinae*)." *Bulletin of the British Museum of Natural History (Zoology)* 43:49–94. (Sulloway's papers are my chief source for the history in Chapter 2.)

———. 1982. "Darwin and His Finches: The Evolution of a Legend." *Journal of the History of Biology* 15:1–53.

———. 1982. "Darwin's Conversion: The *Beagle* Voyage and Its Aftermath." *Journal of the History of Biology* 15:325–96.

———. 1984. "Darwin and the Galápagos." *Biological Journal of the Linnean Society* 21:29–59.

Sutherland, William J. 1990. "Evolution and Fisheries." *Nature* 344:814–15.

———. 1992. "Genes Map the Migratory Route." *Nature* 360:625–26.

Swarth, Harry S. 1934. "The Bird Fauna of the Galápagos Islands in Relation to Species Formation." *Biological Reviews* 9:213–34.

Tax, Sol, ed. 1980. *Evolution after Darwin*. 3 vols. Chicago: University of Chicago Press.

Taylor, Martin. 1990. "Summary of Research, 1988—Present: Population Genetics of Pyrethroid Resistance in *Heliothis*." Unpublished manuscript.

———. 1991. "The Evolution of Resistance to Pyrethroids in Tobacco Budworms." Unpublished lecture notes.

———. 1991. "What Are the Gene(s) Conferring Resistance to Pyrethroids in Tobacco Budworm?" Unpublished lecture notes.

Taylor, Martin, et al. 1993. "Genome Size and Endopolyploidy in Pyrethroid-Resistant and Susceptible Strains of *Heliothis virescens* (Lepidoptera: Noctuidae)." *Journal of Economic Entomology* 86:1030–34.

———. 1993. "Linkage of Pyrethroid Insecticide Resistance to a Sodium Channel Locus in the Tobacco Budworm." *Insect Biochemistry and Molecular Biology* 23:763–75.

Tomasz, Alexander. 1990. "Auxiliary Genes Assisting in the Expression of Methicillin Resistance in *Staphylococcus aureus*." In Novick, Richard P., ed. 1990. *Molecular Biology of the Staphylococci*. New York: V. C. H. Publishers, 565–83.

———. 1990. "New and Complex Strategies of Beta-Lactam Antibiotic Resistance in Pneumococci and Staphylococci." In Ayoub, I. M., et al., eds. 1990. *Microbial Determinants of Virulence and Host Response*. Washington, D.C.: American Society for Microbiology, 345–59.

Toulmin, Stephen, and June Goodfield. 1965. *The Discovery of Time*. New York: Harper & Row.

Vagvolgyi, Joseph, and Maria W. Vagvolgyi. 1990. "Hybridization and Evolution in Darwin's Finches of the Galápagos Islands." *Academia Nazionale Dei Lincei. Atti Dei Convegni Lincei* 85:749–72.

Valen, Leigh Van. 1965. "Morphological Variation and Width of Ecological Niche." *American Naturalist* 99:377–90.

Vitousek, Peter M. 1988. "Diversity and Biological Invasions of Oceanic Islands." In *Biodiversity. See* Wilson, ed. 1988, 181–89.

Vitousek, Peter M., Lloyd L. Loope, and Charles P. Stone. 1987. "Introduced Species in Hawaii: Biological Effects and Opportunities for Ecological Research." *Trends in Ecology and Evolution* 2:224–27.

Vonnegut, Kurt. 1985. *Galapagos*. New York: Delacorte Press.

Wallace, Alfred Russel. 1871. *Contributions to the Theory of Natural Selection*. 2nd ed. New York: Macmillan.

———. 1889. *Darwinism*. New York: Macmillan.

Walsh, Benjamin D. 1867. "The Apple-Worm and the Apple-Maggot." *The American Journal of Horticulture* 2:338–43.

Weidensaul, Scott. 1991. *The Birder's Miscellany*. New York: Simon & Schuster.

Werner, Tracey K. 1988. "Behavioral, Individual Feeding Specializations by *Pinaroloxias inornata*, the Darwin's Finch of Cocos Island, Costa Rica." Dissertation, University of Massachusetts.

Werner, Tracey K., and Thomas W. Sherry. 1987. "Behavioral Feeding Specialization in *Pinaroloxias inornata*, the 'Darwin's Finch' of Cocos Island, Costa Rica." *Proceedings of the National Academy of Sciences* 84:5506–10.

Wiggins, Ira L., and Duncan M. Porter. 1971. *Flora of the Galápagos Islands*. Stanford: Stanford University Press.

Williams, George C. 1966. *Adaptation and Natural Selection*. Princeton: Princeton University Press.

Williams, L. Pearce. 1978. *Album of Science: The Nineteenth Century*. New York: Charles Scribner's Sons.

Wills, Christopher. 1989. *The Wisdom of the Genes*. New York: Basic Books.

Wilson, E. O., ed. 1988. *Biodiversity*. Washington, D.C.: National Academy Press.

———. 1992. *The Diversity of Life*. Cambridge, Mass.: Belknap Press of Harvard University Press.

Wood, Thomas K., and M. C. Keese, 1990. "Host-Plant-Induced Assortative Mating in Enchenopa Treehoppers." *Evolution* 44:619–28.

Wood, Thomas K., K. L. Olmstead, and S. I. Guttman. 1990. "Insect Phenology Mediated by Host-Plant Water Relations." *Evolution* 44:629–36.

Woodruff, R. C., and J. N. Thompson. 1980. "Hybrid Release of Mutator Activity and the Genetic Structure of Natural Populations." *Evolutionary Biology* 12:129–62.

Yang, Suh Y., and James L. Patton. 1981. "Genic Variability and Differentiation in the Galápagos Finches." *Auk* 98:230–42.

Young, Robert M. 1985. *Darwin's Metaphor.* Cambridge, U.K.: Cambridge University Press.

Zhang, Ying, et al. 1992. "The Catalase-Peroxidase Gene and Isoniazid Resistance of *Mycobacterium tuberculosis.*" *Nature* 358:591–93.

Zimmerman, Elwood C. 1960. "Possible Evidence of Rapid Evolution in Hawaiian Moths." *Evolution* 14:137–38.

———. 1971. "Adaptive Radiation in Hawaii with Special Reference to Insects." In Stern, William L., ed. 1971. *Adaptive Aspects of Insular Evolution.* Pullman: Washington State University Press, 32–38.

Acknowledgments

I met the Grants in January 1990, when they were on their way down to the Galápagos. Peter put aside the packing to have lunch with me. No one had ever written about the Grants' work in all their time on Daphne Major (aside from a few squibs in newspapers and magazines) and despite Peter's enormous charm I could tell he was not eager for me to write about them either. In the middle of lunch he sprang up from his chair to pace off the size of their campsite on Daphne. "Not much bigger than this table!" he said cheerfully. Too small for visitors.

After lunch he introduced me to Rosemary and their younger daughter, Thalia, who was joining her parents once again on Daphne. I remember thinking that the three of them might have been picked by central casting to play roles in *The Swiss Family Robinson*. But again along with the lighter-than-air charm there was an equally strong air of reserve, a reluctance to be cast in any roles by anyone. Just before they left for Ecuador, Peter wrote me a note suggesting that I put them out of my mind for a few years, until they had a chance to wrap up their study. He thought I might prefer not to write about a work in progress.

When the Grants got back from the islands that year, they found me sitting in Princeton's Biology Library, reading their team's technical reports: more than 150 scientific papers and several monographs on Darwin's finches. I had also begun gathering other eyewitness reports of evolution in action, a collection that now approaches two thousand technical papers and books. (The ones that contributed most to the final draft of this book are listed in the bibliography.) I was convinced that the Grants' watch on Daphne Major is one of the most remarkable works in progress on this planet, and I had to write about it.

THE GRANTS WARMED TO MY PROJECT (or took pity on me). John Tyler Bonner, who first told me about the Grants and their work, and who made the introduction, encouraged me throughout. I am grateful not only for his help but also his friendship. More than eighty other evolutionary biologists consented to be interviewed. Marty Kreitman, who was then at Princeton, opened his lab to me, and I spent many months there, learning about the study of evolution at the level of DNA. I enjoyed the hospitality of Kreitman and his crew, and wish I could have written more about them here. Many thanks to Hiroshi Akashi, Andrew Berry, Jeffrey Feder, John McDonald, Martin Taylor, and Marta Wayne.

I am also very grateful to many people at the Charles Darwin Foundation in Quito, and at the Charles Darwin Research Station in Puerto Ayora, who helped me make a trip to Daphne Major. Special thanks to David Anderson, one of the Grants' former finch watchers, who gave me a tour of some of their field sites, and showed me the vampire finches of Española.

I could not have written the book without the help of a few other veterans of the long watch. Their names and my debts to them fill this book. Special thanks to Trevor Price for lending me the memoir he wrote on Daphne. It is an amazing adventure story and I have read it many times.

Halfway through the writing I found out that the Grants' daughter Thalia is an artist as well as an ecologist. (She learned both the art and science in the islands.) Since this book is so preoccupied with the successions of generations—generations of finches, and generations of scientists—it seemed fitting to illustrate it with drawings from the books of Charles Darwin, and from the sketchbooks of Thalia Grant.

Peter Boag, John Bonner, John Endler, Lisle Gibbs, Peter and Rosemary Grant, Trevor Price, Laurene Ratcliffe, Dolph Schluter, and Frank Sulloway checked the manuscript for scientific and historical accuracy, and some of them read more than one draft. So did my friends Keith Sandberg and Dick Preston, and Floyd Glenn helped with some of the statistics. My mother, Ponnie, helped me prospect in several libraries; my father, Jerry, gave me a Macintosh. As Darwin wrote to his friend Hooker, "I thank you most sincerely for all your assistance; & whether or no my Book may be wretched you have done your best to make it less wretched."

I am extremely lucky in my editor, Jonathan Segal, and am grateful for all his help with this book. My agent, Victoria Pryor, has been supportive in big ways and little ways, and I am thankful. Ida Giragossian has been a pleasure to work with. I am also grateful to Sonny Mehta, whose enthusiasm for the book has done a great deal for it.

I thank my friends and my family for support of many kinds. My sons, Aaron and Benjamin, have now heard more about Darwin's finches than any other children their age since Nicola and Thalia. My wife, Deborah, read many drafts, and weathered many gusts and squalls.

Working on this book has given me an extraordinary view of evolution in action. For this view, I cannot imagine a better place to stand than the rim of Daphne Major. I want to thank the Grants again, not only for the time and help they have given me, but for the view itself.

Index